North American Game Birds of
Upland and Shoreline

North American Game Birds of Upland and Shoreline

Paul A. Johnsgard

University of Nebraska Press • Lincoln

DEDICATED TO MY STUDENTS
Who, while I thought I was molding them,
have patiently remolded
me

Library of Congress Cataloging in Publication Data

Johnsgard, Paul A
    North American Game Birds of Upland and Shoreline
    Bibliography: p.
    Includes index.
    1. Upland game birds—North America.   2. Shore birds
—North America.   I. Title.
QL681.J64   598.2'973   74-15274
ISBN 0-8032-5811-9 pbk.
ISBN 0-8032-0701-8 cloth

Manufactured in the United States of America

Contents

Illustrations

Figures

Preface

This book is intended as a guide to the more common species of American game birds, fuller than the typical bare-bones field guides but less detailed than a full-scale monographic treatment. The capsule summary for each species includes information on identification of the species and determination of its age and sex, a statement of its geographic range, and a brief review of the species' biology. A list of additional sources of information on these birds will be found in the References at the end of the book.

The decisions on which species should be included in an abbreviated review of the gallinaceous and nonwaterfowl game birds were tied in part both to the relative importance of these species to hunters and to the traditional use of the term *waterfowl*. This term is generally accepted by ornithologists to mean only the ducks, geese, and swans of the family Anatidae, and excludes such equally aquatic birds as loons, grebes, and coots. On the other hand, the term *upland game birds* includes not only the gallinaceous species like grouse, quails, partridges, and pheasants, but also such distantly related birds as pigeons and doves. However, it excludes such marsh-dwelling species as the common snipe, rails, and gallinules, which, although technically not waterfowl, are important game birds in some parts of America.

I finally decided to include all of the native and successfully introduced species of gallinaceous game birds of North America occurring north of Mexico as well as the migratory game birds other than waterfowl that are legally harvested in significant numbers in the United States and Canada. This required the inclusion of two species of shore birds, three species of pigeons and doves, and eight species of cranes, coots, gallinules, and rails. The gallinules and rails are marginal because of their minimal importance as game species, and thus collective rather than separate descriptions are given for them.

All of the line drawings and most of the photographs that are included are my own. In a few instances, photos taken by other persons have been used, and are individually credited. For their use, I would like to acknowledge my thanks to LaVern Duemey, Ken Fink, Bruce Porter, David Allen, Alan Nelson, and Raphael Payne.

Introduction to the Biology of
Game Birds

Although the major purpose of this book is to facilitate the identification of game birds, some basic information is also presented
on the adaptations for survival and reproduction of each of the
kinds of game birds included in the book. Rather than discussing
"kinds" of birds, biologists in general refer to individual or
groups of species. Each species is known by one or more English
vernacular names such as "bobwhite," and by a two-part Lat-
inized scientific name which consists of a general or generic
name and a specific name that form a unique combination. In the
case of the bobwhite, the generic and specific names are *Colinus
virginianus*. Different geographic races are given a third or sub-
specific name, as for example *Colinus virginianus ridgwayi* in the
case of the masked bobwhite of the Sonoran desert region.
Frequently individuals occur that represent "intergrades" between two different subspecies and thus cannot be assigned to a
particular race. Less often, individuals that are intermediate between two different species are found; such hybrids are very rare
in nature but often can be produced in captivity by restricting the
choices for mating.

The reader who may wish to identify a game bird as to its
species, age, and sex must be familiar with certain aspects of bird
anatomy, especially the organization of feather regions on the
body and wings (see figure 28). These areas of feathers first become evident on a young bird after it loses its initial covering of
downy feathers. The first nondowny plumage of young birds is
called the juvenal plumage, and the age class is called the juvenile stage. During this period the first flight feathers of the wings
are grown, including the outer, or primary, feathers and the
inner, or secondary, feathers. Each of these groups of feathers has
a series of shorter feathers that lie immediately in front of them,
which are called coverts. Together with the flight feathers, the
coverts provide important clues for judging the age of game

birds. This is a result of the fact that in many species these feathers are marked or shaped differently in young birds than in adults. Particularly in grouse and quails this is a useful age-estimating device, since in these groups the outer pair of primaries and their associated coverts are carried through the entire first year of life, even though the rest of the flight feathers are molted and replaced with new ones during the first fall and winter. Thus, faded and frayed outer primaries or primary coverts are important indicators of immaturity in these birds.

After the juvenile plumage has been molted, the bird may be called an immature. In many game birds the plumage carried by immatures, which is termed the first-winter plumage, is the one in which they initially breed. This is true of those birds that breed in the year following their hatching. Others, such as cranes, may require three or four years to become reproductively mature; "subadults" of such species are usually not distinguishable on the basis of their plumage from breeding adults.

A second anatomical character that is typical of nearly all of the game birds included in this book is the bursa of Fabricius, an appendixlike structure located near the end of the large intestine (figure 29). It consists of soft lymphoid tissue in which antibodies are produced. Like comparable tissues in mammals, its activity and its size decrease with age, and in adults it may be very small or even absent. In many birds the presence of a bursa can be detected by the use of a probe, without actually opening the body cavity.

The pattern of mating and associated pair bonds varies greatly among game birds. Males in some groups are strongly monogamous and have relatively permanent pair bonds, while others are polygamous, having several female mates, or promiscuous, with no real pair bond being formed at all. Whether one condition or the other prevails is largely a result of ecological factors, and especially whether the presence of the male is needed to help defend and care for the female and her brood. Some general patterns do exist: all of the quail species are strongly monogamous, with the male helping to rear the young, while nearly all of the grouse species are either polygamous or promiscuous.

Whether or not the male helps to defend the eggs or young, he may defend a "territory" from which other males are vigorously excluded. This territory is established by fighting, vocalizations, or various other display behaviors. In some of the grouse such as the sage grouse, pinnated grouse, and sharp-tailed grouse the territory serves only as a mating location, to which females are attracted for fertilization. In others such as the ptarmigans it is much larger, and the female places her nest within its protective boundaries. Thus she avoids harassment from other male ptarmigan and receives protection from potential predators that the male may be able to evict or at least successfully distract. If the initial clutch of eggs is destroyed by predators or otherwise is unsuccessful, a second nesting effort may be made. Such renesting is more common in climates allowing a protracted breeding season than in arctic ones. Although in none of the species considered here is a nest normally shared by two or more females, it sometimes happens that females may deposit an egg or two in a nest other than their own. In some species this "dump nesting" is fairly common, and is the probable precursor to an entirely parasitic mode of nesting like that found in cowbirds.

In most of the species considered here, incubation of the eggs usually does not begin until the clutch is complete. This assures a hatching of all of the young at about the same time, or after the completion of a rather specific incubation period. In a few game birds, such as some rails, some incubation of the eggs laid early in the clutch may occur before the last eggs are laid; this results in a staggered period of hatching. After the young have hatched, they remain in the nest for feeding and parental brooding. This is the situation with the young of doves and pigeons, which are quite helpless when removed from the nest. The semihelpless young of such species are called altricial, and these birds are much more dependent on parental care than are the young of grouse and quails. The down-covered chicks of grouse and quails can run about and forage for themselves within hours after hatching, although they may require periodic brooding by the parent to avoid becoming chilled. Such lively young are called precocial, and in most precocial species the wing feathers begin to grow immediately upon hatching. The initial flights of precocial chicks may occur within a week after hatching. The period to initial flight, called the fledging period, is, together with the incubation period, an important factor in determining the climatic limits within which a species can successfully reproduce.

Greater Prairie Chicken (Territorial Male)

The Hunting and Recreational Value of Upland Game Birds

A national survey made in 1970 by the federal government (*National Survey of Hunting and Fishing*) indicates that over 14 million Americans were then actively engaged in the hunting of big game, small game, or waterfowl. Of this total, some 11.6 million hunted small game, a category that includes not only nonwaterfowl game birds but also small game mammals such as squirrels, rabbits, and a variety of unprotected "varmints." Excluding from this total the roughly 1 million varmint hunters and an unknown number of persons who specialized in hunting small game mammals, it is probable that nearly 10 million Americans regularly hunt one or more of the species described in this book. Such a total would include about 16 percent of the adult male population of the United States. It was further estimated that the small game hunters of America spend nearly a billion dollars annually in pursuit of their sport. On an individual basis, this represents an approximate yearly expenditure of 80 dollars. Clearly, the upland game birds of the United States and Canada constitute an extremely valuable source of hunting recreation for our population, and in some areas hunting represents an important part of the total economy.

Of course not all species of legally hunted game birds are equally important to hunters, since they differ greatly in abundance, accessibility, size, wariness, and suitability for the table, all of which influence their general desirability as game species. Among the grouse and quails, the ruffed grouse and bobwhite quail have the requisite combination of widespread occurrence, adequate body size, and sufficient wariness to make them "sporting" targets. They are both also considered to be excellent table fare, thus adding gourmet value to their basic sporting attractiveness.

Not all the upland game species fill these criteria nearly so well. For example, the sage grouse has the advantage of large size that makes it a fine trophy bird, but the flesh of adults tends to have a sagelike flavor that ranks them as second-rate table birds. The ptarmigans are not only excellent eating but also highly sporting targets, and yet are so inaccessible that they receive little attention from most hunters. The rails are fine table birds, but their flight is so slow and they flush at such close range as to make them too easy a target to satisfy the sporting urges of most hunters. In addition, the sora, which is the most abundant and widespread species, is so small as to be of limited value in terms of the meat it provides for the time and effort expended in its hunt.

Thus the size, and especially the weight, of these species is of considerable interest to sportsmen, both as a measure of the birds' relative trophy values and as an index to their potential for providing a supply of fresh meat for the table. Therefore, a summary of average and maximum weights for adults of the species included in this book is provided in tables 1 and 2. The figures in general represent the largest samples that were located for each species. Since such factors as disease and starvation so strongly influence minimum weights, these figures were not considered worthy of inclusion in this summary. It is quite apparent that, in terms of size alone, the wild turkey is the finest trophy of all the American game birds. The lesser sandhill crane ranks second, followed by the sage grouse. The smaller species of rails and the doves comprise the opposite end of the trophy spectrum.

There is no definite relationship between the average weights of these species and the numbers that are annually harvested by hunters. Indeed, the smaller species tend to be shot in the greatest numbers, probably a reflection of their more common and widespread occurrence as well as the more liberal bag limits associated with these considerations. Estimated total annual harvests of these species from the United States and Canada (table 3) provide an interesting basis for comparison of their relative hunting values. The data in this table indicate that the bobwhite is easily the number one upland game bird in North America, far outstripping its nearest competitor, the ring-necked pheasant. Nearly 50 million individuals of these two species are harvested by hunters every year in the United States and Canada. This represents over 60 percent of the total estimated harvest of 74 million birds of all of the 32 species concerned. These two species also constitute about 70 percent of the overall tonnage of fresh meat represented by these totals, which collectively come to approximately 60 million pounds. All told, the "average" hunter of upland game birds might be expected to kill in the course of a season about seven birds of one or more of these species, which collectively weigh about six pounds. In view of an estimated per capita 80-dollar annual hunting expenditure, this represents an expensive source of meat. Fortunately, most hunters do not calculate their pleasure simply in terms of the amount of meat obtained per dollar of investment.

If hunting is to be defended on any esthetic grounds, it can most convincingly be done on the basis of the potential that it offers for exposing a basically urban-oriented person to the varied beauties of the outdoors, if only for a few days each year. It seems to be a general truism that hunters gradually become more aware of these additional pleasures and the daily or seasonal kill becomes a secondary consideration. Indeed, many persons (the author included) who have had their outdoor interests initially whetted by hunting eventually give it up altogether, and substitute bird-watching, bird photography, hiking, or other nonconsumptive activities for the more competitive and destructive aspects of hunting.

It is extremely difficult to estimate the recreational value of our game birds to persons who think of them as objects to be enjoyed when they are encountered on hikes, or when viewed through binoculars, telescopes, or the finder of a camera rather than through the sights of a firearm. This difficulty is partly the result of a lack of firm statistical data to compare against those of hunting; no license is required to observe or photograph a wild bird. However, the 1970 *National Survey of Hunting and Fishing* estimated that there are some 7.4 million bird-watchers in the United States, plus 4.9 million persons who photograph birds and wildlife, and about 30 million who take nature walks. An added indication of the interest of Americans in wildlife is the fact that in 1970 there were some 18 million visitors to our national wildlife refuges. Although only about 40 percent of these visits were concerned with interpretive aspects of wildlife enjoyment such as bird-watching, it is clear that the proportion of Americans who elect to enjoy wildlife through nonconsumptive methods may be at least as great as or possibly greater than those who prefer to carry a gun when afield.

It has been estimated that, on the basis of data derived from the National Audubon Society's annual Christmas counts, the bobwhite, ruffed grouse, and California quail are the three species of grouse and quails that are most commonly encountered by bird-watchers during the winter months. Thus, at least in terms of relative opportunities for man-bird contacts, they offer the greatest recreational potential for birders at that time of year. At that season, of course, migratory species such as the doves and shore birds are virtually absent from Canada and the United States, and their recreational value would be seriously underestimated. Furthermore, the majority of bird-watchers are more active in late spring and summer, when both the weather conditions and the bird populations may be more favorable for enjoyment by the average person.

One source of information on the relative summer abundance and associated possible contacts between man and birds is provided by the annual breeding bird survey of the Bureau of Sport Fisheries and Wildlife. This survey was initially organized on a major regional basis in 1966, when the states and provinces east of the Mississippi River were surveyed. By 1967, several more westerly states were included in the survey, and in 1968 the first coast-to-coast survey was attempted. By that time nearly 1,200 survey routes, each with 50 stops, had been established, and a fairly representative sampling of the continent's bird populations was obtained. Of the species included in this book, only three have consistently been observed in numbers greater than the total number of surveys (i.e., three). These are the mourning dove, bobwhite, and ring-necked pheasant, which, respectively, averaged 24, 19, and 5 encounters per route in 1968. Species that were never encountered during the three years 1966 through

1968 included the chachalaca, all three ptarmigans, and the harlequin quail, while the spruce grouse was encountered only once. Although such rarely encountered species have limited recreational value for most bird-watchers, they have an opposite, if elusive, "rarity value" that for some bird-watchers represents the ultimate measure for their pastime.

Spruce Grouse (Male)

Table 1
Adult Weights of American Gallinaceous Game Birds

Species	Mean or Range of Means	Maximum Weight	Authority (See References)
Chachalaca			
Male (8 birds)	19.8 oz. (561 gm.)	24.2 oz.	This study
Female (9 birds)	17.7 oz. (502 gm.)	19.1 oz.	
Turkey			
Male (54 birds)	16.3 lb. (7393 gm.)	23.8 lb.	#12 ‡
Female (55 birds)	9.3 lb. (4218 gm.)	12.3 lb.	
Sage grouse			
Male	71–100 oz. (2010–2835 gm.) *	112 oz.	#70
Female	40–54 oz. (1142–1531 gm.) *	54 oz.	
Blue grouse			
Male	40–45 oz. (1150–1275 gm.)	50 oz.	#30 †
Female	30–32 oz. (850–900 gm.)	44 oz.	
Spruce grouse			
Male (14 birds)	17.7 oz. (501 gm.)	22 oz.	#77
Female (15 birds)	17.5 oz. (497 gm.)	21 oz.	
Willow ptarmigan			
Male	19–25 oz. (535–696 gm.) *	28 oz.	#13
Female	19–23 oz. (525–652 gm.) *	26 oz.	#6
Rock ptarmigan			
Male	16–19 oz. (466–536 gm.) *	21 oz.	#6
Female	15–18 oz. (427–515 gm.) *	20 oz.	#9
White-tailed ptarmigan			
Male (24 birds)	11.4 oz. (323 gm.)	15.2 oz.	#8
Female (14 birds)	11.5 oz. (329 gm.)	17.5 oz.	#7
Ruffed grouse			
Male	21.5–23.3 oz. (604–654 gm.) *	27 oz.	#12 ‡
Female	17.9–20.9 oz. (500–586 gm.) *	24 oz.	#65
Greater prairie chicken			
Male (22 birds)	35 oz. (992 gm.)	48 oz.	#12
Female (16 birds)	29 oz. (770 gm.)	36 oz.	
Attwater prairie chicken			
Male (10 birds)	33.1 oz. (938 gm.)	40 oz.	#49
Female (6 birds)	25.7 oz. (731 gm.)	28 oz.	
Lesser prairie chicken			
Male (20 birds)	27.6 oz. (780 gm.)	31.5 oz.	#49
Female (5 birds)	25.5 oz. (722 gm.)	27.5 oz.	
Sharp-tailed grouse			
Male (236 birds)	33 oz. (951 gm.)	43 oz.	#12 ‡
Female (247 birds)	29 oz. (815 gm.)	37 oz.	

Species	Mean or Range of Means	Maximum Weight	Authority (See References)
Mountain quail			
Male (30 birds)	8.2 oz. (235 gm.)	10.3 oz.	#1
Female (24 birds)	8.2 oz. (230 gm.)	10.0 oz.	#10
Gambel quail			
Male (390 birds)	5.7 oz. (161 gm.)	6.6 oz.	#3
Female (337 birds)	5.6 oz. (156 gm.)	6.7 oz.	
California quail			
Male (418 birds)	6.2 oz. (176 gm.)	7.3 oz.	#12 ‡
Female (272 birds)	6.0 oz. (162 gm.)	7.3 oz.	
Scaled quail			
Male (143 birds)	6.7 oz. (191 gm.)	8.2 oz.	#3
Female (132 birds)	6.2 oz. (177 gm.)	7.7 oz.	#12 ‡
Bobwhite (eastern U.S.)			
Male (899 birds)	6.1 oz. (173 gm.)	9.0 oz.	#12
Female (692 birds)	6.0 oz. (170 gm.)		
Harlequin quail			
Male (45 birds)	6.9 oz. (195 gm.)	7.9 oz.	#42
Female (22 birds)	6.2 oz. (176 gm.)	7.1 oz.	
Gray partridge			
Male (87 birds)	14 oz. (396 gm.)	16.0 oz.	#12 ‡
Female (57 birds)	13.7 oz. (379 gm.)	15.3 oz.	
Chukar partridge			
Male (44 birds)	19.6 oz. (557 gm.)	22.3 oz.	#37
Female (50 birds)	15.7 oz. (444 gm.)	18.5 oz.	
Ringed-necked pheasant			
Male (6378 birds)	46.4 oz. (1315 gm.)	65.6 oz.	#12 ‡
Female (759 birds)	33.6 oz. (952 gm.)	51.2 oz.	

* Mean weights of these species vary considerably.
† Reported in graphic form, points interpolated.
‡ Reported as fractions of pounds by authors.

Table 2

Weights of Other Game Birds

Species	Mean or Range of Means	Maximum Weight	Authority (See References)
Lesser sandhill crane			
Adult (178 birds)	7.0 lb. (3175 gm.)	9.25 lb.	#73
Immature (52 birds)	6.25 lb. (2835 gm.)	———	
King rail			
Male (9 birds)	14.7 oz. (416 gm.)	17.3 oz.	#58 *
Female (9 birds)	10.8 oz. (306 gm.)	11.5 oz.	
Clapper rail			
Male (15 birds)	11.3 oz. (322 gm.)	12.3 oz.	#58 *
Female (7 birds)	9.6 oz. (271 gm.)	9.7 oz.	
Virginia rail			
Male (3 birds)	3.8 oz. (108 gm.)	4.3 oz.	#12 †
Female (2 birds)	2.9 oz. (82 gm.)	3.3 oz.	
Sora			
Male (4 birds)	2.1 oz. (59 gm.)	3.0 oz.	#12 †
Female (4 birds)	2.1 oz. (59 gm.)	2.8 oz.	
Purple gallinule			
Both sexes (38 birds)	7.6 oz. (215 gm.)	9.3 oz.	#12 †
Common gallinule			
Male (10 birds)	10.2 oz. (288 gm.)	11.9 oz.	#16 *
Female (10 birds)	8.5 oz. (241 gm.)	10.4 oz.	
American coot			
Male (17 birds)	20.2 oz. (573 gm.)	26.0 oz.	#12 †
Female (12 birds)	17.4 oz. (493 gm.)	21.0 oz.	
Woodcock			
Male (390 birds)	6.2 oz. (176 gm.)	7.8 oz.	#12 †
Female (313 birds)	7.7 oz. (218 gm.)	9.8 oz.	
Common snipe			
Male (15 birds)	4.5 oz. (127 gm.)	5.5 oz.	#12 †
Female (14 birds)	4.1 oz. (116 gm.)	5.5 oz.	
Band-tailed pigeon			
Male (194 birds)	14.2 oz. (404 gm.)	18.0 oz.	#26 *
Female (139 birds)	13.6 oz. (386 gm.)	16.6 oz.	
Mourning dove			
Male (164 birds)	4.6 oz. (130 gm.)	6.0 oz.	#12 †
Female (80 birds)	4.4 oz. (125 gm.)	5.5 oz.	
White-winged dove			
Male (9 birds)	5.5 oz. (156 gm.)	6.4 oz.	#12 †
Female (2 birds)	5.4 oz. (153 gm.)	5.8 oz.	

* Reported as grams by authors.
† Reported as fractions of pounds by authors.

Table 3

Estimated Annual Harvest, Gallinaceous Game Birds and Migratory Nonwaterfowl

	United States	Canadian Provinces	Total
Wild turkey [1]	84,000	0	84,000
Sage grouse [2]	250,000	few	250,000
Blue grouse [2]	240,000	130,000	370,000
Spruce grouse [2]	140,000	300,000	440,000
Ptarmigans [2]	100,000	200,000	300,000
Ruffed grouse [2]	2,700,000	1,000,000	3,700,000
Prairie chicken [2]	85,000	0	85,000
Sharp-tailed grouse [2]	255,000	200,000	455,000
Mountain quail [2]	375,000	few	375,000
Scaled quail [2]	3,600,000	0	3,600,000
Gambel quail [2]	1,300,000	0	1,300,000
California quail [2]	2,200,000	few	2,200,000
Bobwhite [2]	35,000,000	few	35,000,000
Harlequin quail [2]	6,000	0	6,000
Chukar partridge [2]	650,000	8,000	658,000
Gray partridge [2]	400,000	250,000	650,000
Ring-necked pheasant [2]	12,000,000	?	12,000,000+
Lesser sandhill crane [2]	3,000	4,200	7,200
Coot [3]	885,000	69,000	954,000
Gallinules [4]	19,000	2,000	21,000
Sora rail [4]	31,000	?	31,000+
Other rails [4]	60,000	?	60,000+
Woodcock [4]	452,000	107,000	559,000
Snipe [4]	333,000	103,000	463,000
Mourning dove [4]	9,822,000	4,000	9,826,000
White-winged dove [4]	438,000	0	438,000
Band-tailed pigeon [4]	217,000	11,000	228,000

[1] Based on six years' data in U.S. Fish and Wildlife Service's *Wildlife Leaflets* 446, 454, 461, 470, 473, and 477.
[2] Based on author's estimates from data of individual states and provinces.
[3] Based on Fish and Wildlife Service's *Waterfowl Status Reports,* 1965–70, and Canadian Wildlife Service's *Progress Notes* 16, 22, and 28.
[4] Based on five years' data in Fish and Wildlife Service's *Special Scientific Report (Wildlife)* no. 142, and three years' data in Canadian Wildlife Service's *Progress Notes* 16, 22, and 28.

Glossary

Anatidae: that family of birds which includes ducks, geese, and swans.

Arboreal: adapted to tree living.

Charadriiformes: that order of birds which includes the true shore birds (Charadriidae) and their relatives.

Circumpolar: referring to a distribution that encompasses the polar land masses of both eastern and western hemispheres.

Columbiformes: that order of birds which includes pigeons and doves (Columbidae) and their relatives.

Composite: referring to plants of the Compositae family, such as asters.

Cracid: referring to birds of the Cracidae family, including chachalacas, guans, and curassows.

Crustacean: an animal of the class Crustacea, which includes crayfish, crabs, and their relatives.

Deciduous: among plants, those species that shed their leaves annually.

Fledging period: the period in birds between hatching and initial flight.

Forb: any herbaceous plant except grasses and grasslike species.

Galliformes: that order of birds which includes pheasants, partridges, quails, and their relatives.

Gallinaceous: belonging to the bird order Galliformes.

Gruiformes: that order of birds which includes cranes (Gruidae) and their relatives.

Herbaceous: nonwoody plants that die at the end of a growing season or after producing seeds.

Immature: that age class in birds following the juvenile period, typically the first winter after hatching.

Incubation period: the period required between the start of incubation and hatching.

Invertebrate: any animal lacking a spinal column, including all of the "lower" animals.

Juvenal: pertaining to feathers of the juvenile age class.

Juvenile: that age class in birds in which initial flight (fledging) occurs, and which precedes the immature stage.

Lateral: toward the side of a structure, as opposed the middle.

Leguminous: pertaining to plants of the family Leguminosae, such as peas, beans and other legumes.

Mast: acorns and similar nutlike fruit from trees.

Nuptial plumage: the plumage in which breeding occurs among birds.

Pinnated: having pinnae (elongated feathers) in the neck region.

Race: a geographic subdivision of a species, also called a subspecies.

Rallidae: that family of birds which includes rails, coots, and gallinules.

Resident: a nonmigratory species or population of birds.

Sternum: the breastbone of birds, which usually is deeply keeled.

Subadult: an age class in birds prior to attainment of reproductive maturity, found in species that do not breed the first year after hatching.

Tinamous: fowllike birds of the family Tinamidae, native to Central and South America.

Vermiculations: fine, wavy pigmentation patterns on feathers that vaguely resemble worm tracks.

Grouse, Quails, Partridges, Pheasants, and Ptarmigans

(Order Galliformes)

Chachalaca

Ortalis vetula (Wagler) 1830

Other vernacular names. Northern chachalaca, plain chachalaca.

Range. Resident from the lower Rio Grande Valley, Texas, southward through Nuevo León and Tamaulipas along the Gulf of Mexico and Caribbean drainage to Nicaragua and Costa Rica. Introduced and established on Sapelo and Blackbeard islands, Georgia. Populations of the larger chachalaca from western Mexico are now believed to represent a separate species (*O. polioce-phala*), as are the white-bellied chachalacas (*leucogastra*), which range from southern Chiapas southward through El Salvador to northern Nicaragua. For the most part the ranges of these three species do not overlap appreciably.

Identification. Adults, 20–24 inches long. Chachalacas are a nearly uniform olive drab color except for the white-tipped tail and a pale abdomen and under tail coverts. A bare area of bluish skin is present in front of the eye, and the bare throat is grayish flesh except in spring, when it is reddish in males. Otherwise there are no apparent external differences between the sexes.

Field Marks. The chachalaca is an olive-colored chickenlike bird with a long, white-tipped tail which lives in dense woodland thickets and is only infrequently seen. It is more inclined to perch than are the native American quails, and is far more often heard calling at dawn and dusk than it is seen. At such times a chorus of distinctively cadenced *cha-cha-lac* calls emanate from the treetops with regularity. A near-synchronous duet by pairs is common, so that an apparent four-noted *cha-cha-lac-ca* call is produced. Reportedly, the other two Mexican species of *Ortalis* actually do have four-syllable calls. Chachalacas seem to fly with some difficulty, but are adept at running on the ground or jumping from branch to branch in the trees. They slightly resemble tinamous in the shape of their head and bill, but the unusually long tail is distinctive. Related species of cracids that occur in the same area of Mexico as this species are the great curassow (*Crax rubra*) and the crested guan (*Penelope purpurascens*). Both of these species are much larger and more distinctly crested than chachalacas.

Age and sex criteria. *Females* average slightly smaller than males, but plumage measurements overlap considerably. The trachea of the adult male loops downward over the breast muscle almost to the end of the sternum, whereas that of the female is normally developed. It is usually possible to detect this without skinning the bird. Judging from birds raised in captivity, this difference is detectable by the time the birds are 5 or 6 months of age.

Immature males lack the tracheal loop mentioned above, although the age at which the adult condition is attained has not been accurately established. Since the chachalaca has a complete postjuvenal molt of all its wing feathers, the occurrence of outer frayed primaries cannot be used as an indication of immaturity. However, it is reported that immature chachalacas do differ from

FIGURE 1. Current distribution of the chachalaca.

adults in a number of plumage traits, including a tail that is tipped with ashy brown rather than white, a more brownish body coloration, and wing feathers tipped and marbled with cinnamon coloration.

Habitat and foods. In southern Texas and most of its Mexican range, this chachalaca is found in areas of brush or woody thickets. Farther south it also occurs in openings of rain forests and in some drier habitats. Wooded stream beds, such as those with dense growths of ebony, hackberry, and mesquite, and thick, shrubby undergrowths are preferred habitats. From such vegetation the birds obtain their primary foods of berries; green, leafy material; and fresh shoots. They also eat ground-dwelling insects to some degree, although they spend most of their time in trees.

Social behavior. The most conspicuous feature of chachalacas—indeed the most obvious sign of their occurrence in an area—is the chorus that the birds perform every morning and evening. Both sexes join in, and the appreciably lower pitch of the calls of the older males distinguish them from the notes of females and younger males. Flocks may number twenty or more birds, and the din produced by such a number of calling birds is truly impressive. As the morning chorus dies down, the birds turn to foraging, but on moonlit nights the evening chorus may persist well after dark. Calling is most evident during the mating season, but ends as the pairs separate from the flocks and begin nesting activities.

Reproductive biology. Chachalacas build small, frail nests in trees, bushes, vines, or, rarely, on the ground. They are often built in the fork of a tree, and may be composed of sticks, leaves, and Spanish moss. The male remains with the female, and usually assists with or even takes the major role in building the nest. The clutch normally numbers 3 eggs, but dump nests may contain many more. Incubation is by the female, and lasts from 22 to 25 days. Within only a few days after hatching the chicks can fly short distances, and thereafter they become progressively more arboreal. They are initially fed by regurgitation of fruit by the parents, mainly the male.

Turkey
Meleagris gallopavo Linnaeus 1758

Other vernacular names. Gobbler, tom, wild turkey.

Range. The native range currently extends from Arizona, Colorado, extreme southern Kansas, Missouri, Kentucky, West Virginia, Pennsylvania, and southern New York south to Michoacán, Veracruz, and the Gulf Coast of the U.S. eastward to Florida. Formerly north to eastern Nebraska, southeastern South Dakota, Iowa, southern Wisconsin, central Michigan, southern Ontario, Massachusetts, southern Vermont, southern New Hampshire, and southwestern Maine. Recently successfully introduced or reintroduced into California, Indiana, Iowa, Kansas, Massachusetts, Michigan, Montana, Nebraska, North Dakota, Ohio, Oregon, South Dakota, Utah, Washington, Wisconsin, Wyoming, and Vermont.

Identification. Adults, 34 inches long (females) to 48 inches long (males). The wild turkey is virtually identical to the familiar "bronze" barnyard variety, and is easily identified to species by the combination of its very large size and naked head. Wild birds are slimmer than their domesticated counterparts, and the bronze domestic turkey differs from the wild U.S. forms (but not the Mexican ones) in having the tail and upper tail coverts tipped with white.

Field marks. The extremely large size, fan-shaped tail, naked bluish head, and generally bronze plumage coloration should distinguish wild turkeys from all birds except domesticated varieties of this species. The familiar "gobbling" call of males is identical to that of the domestic form, and can often be heard in spring where wild turkeys occur.

Age and sex criteria. *Females* may be distinguished from adult males by their smaller size and their lack of a beard (although a beard up to 3 inches long is sometimes present), a wattle, a caruncled (wrinkled and fleshy) forehead, or a pencillike enlargement ("leader") behind the bill. The back and breast feathers of females are tipped with yellowish brown or whitish coloration, while those of males are black-tipped, which gives females a generally lighter appearance than males. Females normally lack spurs on their legs, and their tracks usually measure less than 4.25 inches from the claw of the middle toe to the posterior part of the "heel." The distance from the ventral base of the hind toe to the bend of the actual heel (where feathering begins) is about 4.5 inches in females and 6 inches in males. The typical stride of females (of the eastern race) is from 7.3 to 10.9 inches, while in males it is from 11.9 to 14.2 inches. The droppings of females usually range up to 8 mm. (¼ inch) in diameter, while those of yearling males are up to 10 mm. (⅜ inch) and of adult males up to 15 mm. (⅝ inch).

Immatures retain sharply pointed and indistinctly barred outer primary feathers until they are 12–14 months old. Birds of the year also have two central tail feathers that are longer than those toward the outside. Age-related differences in the greater wing coverts are often more obvious than these wing and tail differ-

FIGURE 2. Current distribution of the turkey.

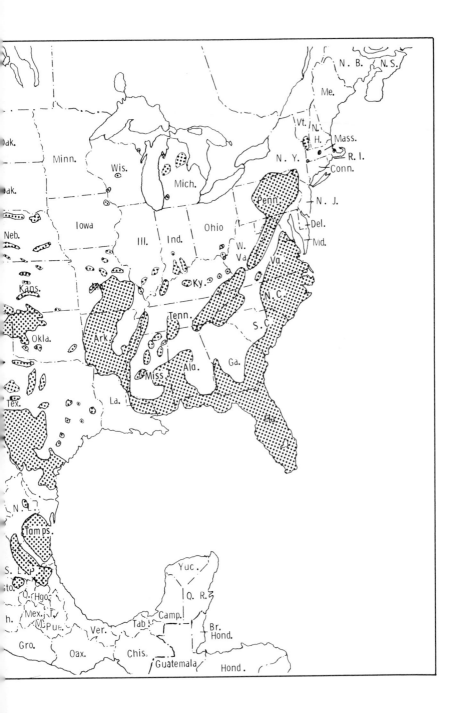

ences. The greater coverts of adults are uniformly dark and glossy, and are generally longer and more symmetrical than those of young birds. The coverts of immatures are increasingly shorter, browner, and less glossy toward the outside, except for the two most distal coverts, which are longer, darker, and glossier than the more proximal ones.

Habitat and foods. The several races of wild turkeys native to the United States occupy habitats that range from the wooded swamps of the eastern and southeastern states to the sparsely wooded flatlands and river bottoms of the southern Great Plains and the coniferous forests of the western mountains. Open hardwood forests containing mature mast-bearing trees such as oaks are preferred by the eastern race, especially if they have occasional open spaces. The southern Rio Grande race is typically found in more arid and grass-dominated habitats, but open-topped roosting trees, a water supply, and succulent vegetation are important components of its habitat. The Merriam race of the western states is most often associated with ponderosa or subalpine pines, arid-adapted oaks, junipers, running water, and a rugged topography. In the eastern states, acorns and other nuts are major foods, but to a limited degree turkeys will also eat buds in the manner of grouse. In the Southwest, grasses and seeds are a more important part of the diet than is mast. During the summer, insects, and especially grasshoppers, are regularly eaten by young birds, and to some extent also by adults.

Social behavior. The elaborate and familiar strutting displays of the domestic turkey provide a good idea of the social behavior patterns of wild turkeys, which are likewise polygamous. In late winter the flocks of adult males begin to break up, and the "gobbling" season begins. Gobbling males tend to avoid one another, and by their calls and strutting postures attract adult females. Display areas, or "strutting grounds," may be occupied by several rival males, but these birds are usually separated by several hundred yards. Males may also strut in more than one area, and in each typically defend a territory of only a few acres. Definite territorial boundaries may not be evident, but gobblers drive away any rivals from the hens that they manage to attract. A harem averaging 5 or 6 females, rarely as many as 14, is thus gathered. These females visit the gobbler almost daily during the egg-laying period, after which they leave the male and begin incubation.

Reproductive biology. Females usually select nesting sites that are located near the strutting ground as well as close to a source of water. The nest is normally well concealed by low vegetation and also situated so as to have at least one ready escape route. Eggs are laid at the average rate of 2 every 3 days, and a clutch of 10–13 is typical. Incubation requires about 28 days, and is entirely the female's responsibility. She also undertakes the care of the young, which soon begin to feed actively on their own. Within a week the chicks are able to make short flights and thereafter begin to roost in trees, becoming much less vulnerable to ground predators. Sometimes 2 or more hens with broods will join company, and family groups are likely to remain intact until the next breeding season. Although a few gobblers may join such groups, the adult males usually remain in separate small flocks until the start of the next breeding season.

Sage Grouse

Centrocercus urophasianus (Bonaparte) 1827

Other vernacular names. Sage hen, sage cock, sage chicken.

Range. From central Washington, southern Idaho, Montana, southeastern Alberta, southern Saskatchewan, and western North Dakota south to eastern California, Nevada, Utah, western Colorado, and southeastern Wyoming.

Identification. Adults, 19–23 inches long (females), 26–30 inches long (males). The large size and sagebrush habitat of this species make it unique among grouse. Both sexes have narrow, pointed tails, feathering to the base of the toes, and a variegated pattern of grayish brown, buff, and black on the upper parts of the body, with paler flanks and a diffuse black abdominal pattern. In addition, males have blackish brown on the throat, narrowly separated by white from a dark V-shaped pattern on the neck. Their white breast feathers conceal two large areas of olive skin, the gular sacs. Behind the margins of these gular sacs are a group of short white feathers with stiffened shafts, which grade into longer and softer white feathers and finally into a number of long, black hairlike feathers that are erected during display. Males also have rather inconspicuous yellow eye-combs that are enlarged during display. Females lack all these specialized structures but otherwise generally resemble males. Their throats are buff with blackish markings, and their lower throats and breasts are barred with blackish brown.

Field marks. The combination of sage habitat, large body size, pointed tail, and black abdomen is adequate for certain identification. Males take flight with some difficulty and fly with their bodies held horizontally; females take off more readily and while in flight their bodies dip alternately from side to side. When the bird is in flight the white under wing coverts contrast strongly with the blackish abdomen.

Age and sex criteria. *Females* may readily be distinguished from adult males by their smaller weights and measurements, by the absence of black on the upper throat, and by the fact that the white on their under tail coverts extends from the tip part way down the shaft of the feather.

Immatures (under one year old) resemble females but are paler. Their outer primaries are more pointed and mottled than the others; and their coverts are narrowly pointed and instead of being unmottled dark gray are marked with brown and white and have white tips. Immatures also have light yellowish green toes, while those of adults are dark green. Males do not usually achieve their full breeding condition their first year; subadult males have narrower white breast bands than do adults. The tail feathers of immature males are also blunter and are tipped with white. During their first fall immature birds have bursa depths in excess of 10 mm. (⅜ inch), whereas adults have maximum bursa depths of 7 mm., or about ¼ inch.

Habitat and foods. Throughout its range, the sage grouse is in-

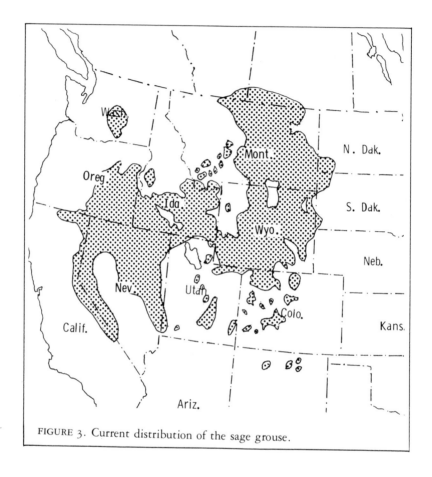

FIGURE 3. Current distribution of the sage grouse.

timately associated with various species of sagebrush the year round. Sage provides winter escape cover and nearly all of the species' food. In spring the males return to their established strutting grounds, which are usually in low, open stands of sage on knolls or ridges. Most nests are hidden under sagebrush, and the young are reared in cover that comprises a diversity of plant forms, including sage species. Not surprisingly, sage is a major item of food for sage grouse through the winter and well into spring, when they begin to eat a variety of herbaceous legumes, broad-leaved weeds, and grasses. Newly hatched birds feed heavily on insects their first few weeks but soon add a variety of nonwoody weeds such as dandelions to their diet. Somewhat later sage becomes an important food. During late summer and fall, adults also eat large amounts of leafy material in the form of legumes and herbaceous broad-leaved plants, but gradually return to a sage diet as these food sources decline.

Social behavior. The most famous aspect of the sage grouse's social behavior is their "strutting" behavior—their formation of unusually large display congregations. The number of males that gather on traditional strutting grounds varies greatly, but often exceeds a hundred or more in areas of dense population. As in other grouse that display socially (pinnated grouse and sharptailed grouse), these groups are highly structured, with dominant males ("master cocks") occupying small but central territories. The territories of the progressively less dominant males are organized in a series of generally concentric patterns reflecting the birds' abilities to attain and hold territories against their rivals. The strutting display of males is marked by a complex sequence of stepping, wing-brushing movements, and four increasing inflations of the esophagus with expansion of the two olive-colored skin patches. These movements produce a distinctive combination of squeaking sounds produced by the wings, plopping noises generated by the air sacs, and a weak vocal *wa-um-poo* sound that carries only a short distance. Females are attracted to the groups of displaying males and invariably move toward the territory of the most dominant male, who is thus responsible for fertilizing the majority of the females in the area. Apart from this brief contact between the sexes, the males have little contact with the females until they merge into fall flocks after brood rearing has been completed.

Reproductive biology. After fertilization has been accomplished, the hen apparently leaves the strutting ground for nesting. Although it is not yet certain, probably only one copulation is needed to achieve fertilization, and egg laying may begin almost immediately thereafter. About 10 days are needed to lay the clutch of 8 or so eggs, and another 25–27 days are required for incubation. There is evidently a fairly high rate of nest desertion and nest destruction by predators, principally coyotes, ground squirrels, badgers, and perhaps magpies and ravens. After hatching, females gradually move their broods to places where food is plentiful, usually in relatively moist areas such as hay meadows, river bottom lands, or irrigated areas. After 10–12 weeks the young are fairly independent of their mother. There is a gradual movement away from the breeding grounds toward wintering areas, with immature females being the first to leave and adult males the last. Wintering grounds are usually at considerably lower elevations than nesting areas and sometimes 50 or more miles distant.

Wild Turkey, Male Displaying

Sage Grouse, Female

Sage Grouse, Males

Sage Grouse, Males

Sooty Blue Grouse, Male

Richardson Blue Grouse, Male

Canada Spruce Grouse, Male

Franklin Spruce Grouse, Male Ken Fink

Willow Ptarmigan, Male in Spring

Willow Ptarmigan, Female Bruce Porter

Rock Ptarmigan, Male

Rock Ptarmigan, Female Ken Fink

White-tailed Ptarmigan, Male

White-tailed Ptarmigan, Male and Female

Ruffed Grouse, Female

Ruffed Grouse, Male

Greater Prairie Chicken, Male

Greater Prairie Chicken, Male and Female

Lesser Prairie Chicken, Male

Sharp-tailed Grouse, Male Alan Smith

Scaled Quail, Male

Gambel Quail, Male and Female

California Quail, Male

Bobwhite, Adults

Harlequin Quail, Male

Harlequin Quail, Female

Chukar Partridge, Male

Gray Partridge, Male

Ring-necked Pheasant, Male

Lesser Sandhill Crane, Subadult

Coot, Adult

Mourning Dove, Adult LaVern Duemey

Blue Grouse

Dendragapus obscurus (Say) 1823

Other vernacular names. Dusky grouse, fool hen, gray grouse, hooter, mountain grouse, pine grouse, pine hen, sooty grouse.

Range. From southeastern Alaska, southern Yukon, southwestern Mackenzie, and western Alberta southward along the offshore islands to Vancouver and along the coast to northern California, and in the mountains to southern California, northern and eastern Arizona, and west-central New Mexico.

Identification. Adults, 17.2–18.8 inches long (females), 18.5–22.5 inches long (males). This is the largest of the coniferous-forest grouse of the western states and provinces. Sexes differ somewhat in coloration, but both have long, squared, and relatively unbarred tails (pale grayish tips usually occur in both sexes of most races). The upper parts of males are mostly grayish or slate-colored, extensively vermiculated, and mottled with brown and black markings; and the upper wing surfaces are more distinctly brown. White markings are present on the flanks and under tail coverts, and feathering extends to the base of the middle toe. The bare skin over the eyes of males is yellow to yellow orange, and the bare neck skin exposed during sexual display varies from a deep yellow and deeply wrinkled condition to purplish and somewhat smoother. Females have smaller areas of bare skin and are generally browner overall, with barring or mottling on the head, scapulars, chest, and flanks.

Field marks. Blue grouse are likely to be confused only with the similar but smaller spruce grouse, the ranges of which overlap in the Pacific Northwest. Male blue grouse lack the definite black breast patch of male spruce grouse. Female blue grouse have relatively unbarred, grayish under parts, compared with the spruce grouse's white under parts with conspicuous blackish barring. A series of 5–7 low, hooting notes is frequently uttered by territorial males in spring.

Age and sex criteria. *Females* may be recognized by barring on the top of the head, nape, and back, which is lacking in adult males, and by the bases of the neck feathers around the air sacs, which are grayish brown rather than white. The sex of adults may be determined from the wings alone: females have a more extensively mottled brownish pattern on their anteriormost lesser wing coverts; in males these feathers are gray, with little or no mottling.

Immatures (in first-winter plumage) may be recognized by one or more of the following criteria: the outer 2 primaries are relatively frayed and more pointed as well as lighter and more spotted than the inner ones; the outer tail feathers are narrow and more rounded (up to ⅛th inch wide at ½ inch below the tip, as opposed to at least 1¼ inch wide in adults); and the tail is shorter than in adults. Immatures of both sexes generally resemble adult females but can usually be recognized by their pale buff or white breasts, the absence of a gray area on the belly, and the absence of a gray bar at the end of the tail.

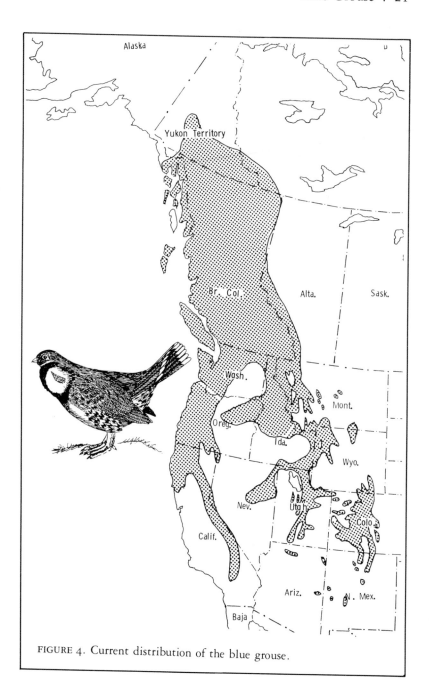

FIGURE 4. Current distribution of the blue grouse.

Habitat and foods. Over its entire range the blue grouse is associated with the occurrence of true firs and Douglas fir, perhaps as a reflection of its winter food needs—needles of these trees and hemlock, as well as buds, twigs, seeds, and pollen cones of the same tree species. As winter ends, the birds move to lower altitudes into dryer and generally more open cover, often in forest-edge habitats. Females generally nest in habitats that provide opportunities for the young to forage on insects and the like. Often these are dry, wooded areas but sometimes they are far from timber. Initially the young feed almost entirely on animal material, but later they begin to eat a variety of berries. By the time they are adults, over 90 percent of their food is plant material and, typically, nearly two-thirds of it is conifer needles.

Social behavior. In early spring males begin to establish "hooting territories," usually where there is a combination of fairly heavy cover for escape and relatively open vegetation for display sites. Their territorial proclamation is a series of low-pitched owllike calls that may be relatively weak or fairly strong and penetrating. During hooting the male partially lifts and spreads his tail and opens the feathers of his neck to expose an oval skin patch that is surrounded by white-based feathers. In some races this skin area is smooth and purplish, while in others it is yellowish and heavily wrinkled. Bare eye-combs are also present and when engorged with blood may vary from bright yellow to a livid red. Most hooting occurs in early morning and again in the evening, and serves to attract females that are ready to begin nesting. The presence of a female produces an intense strutting display by the male which precedes copulation. The male subsequently plays no further role in assuring the female's nesting success.

Reproductive biology. After mating, the female begins nesting. Evidently nearly all females, including yearlings, attempt nesting. Nests are built in quite varied locations, but often are under old logs or among the roots of fallen trees in fairly open timber. Eggs are laid at the rate of 2 every 3 days, and 6–8 eggs constitute the normal clutch. The incubation period is 26 days. After hatching, the chicks rapidly become independent. They are able to fly short distances by 6 or 7 days of age, and by 2 weeks they can fly about 200 feet. Initially the broods use cover that is largely grasses and nonwoody weeds; but as these more open habitats dry up, the birds move into deciduous thickets for the remainder of the brooding period. Gradually the broods break up and the young birds disperse singly or in small groups, slowly working their way upward toward the wintering ranges.

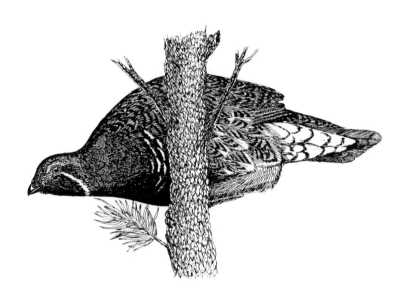

Spruce Grouse
Canachites canadensis (Linnaeus) 1758

Other vernacular names. Canada grouse, cedar partridge, fool hen, Franklin grouse, spruce partridge, swamp partridge, wood grouse.

Range. From central Alaska, Yukon, Mackenzie, northern Alberta, Saskatchewan, Manitoba, Ontario, Quebec, Labrador, and Cape Breton Island south to northeastern Oregon, central Idaho, western Montana, northwestern Wyoming, Manitoba, northern Minnesota, northern Wisconsin, Michigan, southern Ontario, northern New York, northern Vermont, northern New Hampshire, Maine, New Brunswick, and Nova Scotia.

Identification. Adults, 15–17 inches long. This species is associated with coniferous forest throughout its range. The sexes are quite different in coloration, but both have brown or blackish tail feathers that are unbarred and are narrowly tipped with white (western races) or have a broad pale brownish band at the tip. The upper tail coverts are relatively long (extending to about half the length of the exposed tail) and are either broadly tipped with white (western races) or tipped more narrowly with grayish white. The under tail coverts of males are black with white tips and of females are barred. Feathering extends to the base of the toes. Males are generally marked with gray and black above, with a black throat and a well-defined black breast patch that is bordered with white-tipped feathers. The abdomen is mostly blackish, tipped with tawny to white markings that become more conspicuous toward the tail. The bare skin above the eyes of males is scarlet; they have no bare skin on the neck. The females are extensively barred on the head and under parts with black, gray, and yellowish buff in varying proportions; the sides are predominantly ocher and the under parts are mostly white.

Field marks. In the eastern states and provinces spruce grouse are likely to be confused only with the ruffed grouse, from which the spruce grouse can be readily discriminated by its unbarred tail and the presence of a lighter tip rather than a darker band toward the tip of the tail. The conspicuous black and white markings on the under parts of males distinguish spruce grouse from blue grouse, and the predominantly white under parts of females help to distinguish them from the generally similar female blue grouse.

Age and sex criteria. *Females* can be distinguished from adult males by their tawny to whitish throats and breasts, barred with dark brown (these areas are black or black tipped with white in males). Accurate determination of sex in most races is possible by using either the breast feathers (males' breast feathers are black tipped with white, those of females are barred with brown) or by the tail feathers (males have black feathers, tipped and lightly flecked with brown; females have black or grayish brown, heavily barred with brown). In the western (Franklin) race the breast coloration is the same as above, but the tails of females are barred or

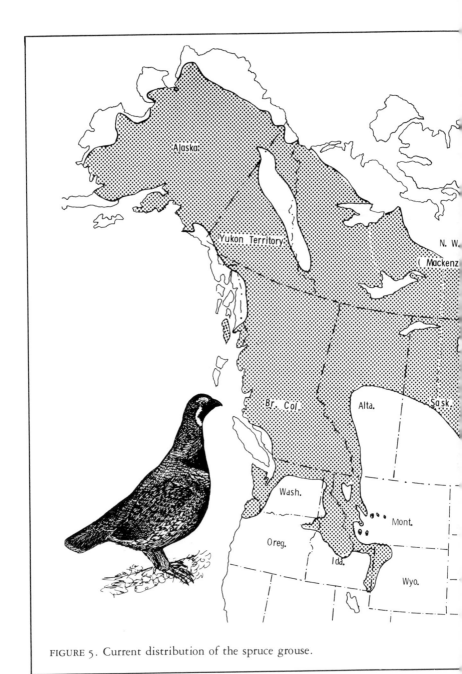

FIGURE 5. Current distribution of the spruce grouse.

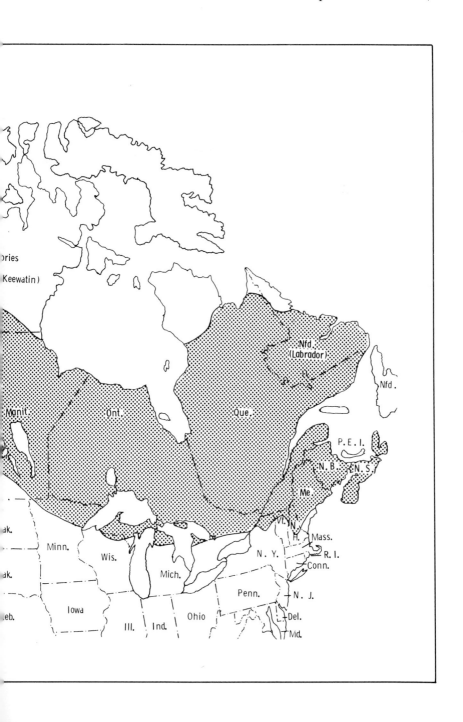

ries

(Keewatin)

Nfd.
(Labrador)

Nfd.

Manit

Ont.

Que.

P.E.I.

N.B.

N.S.

Me.

Vt. N.
H.

Mass.

ak.

Minn.

Wis.

N.Y.

R.I.
Conn.

ak.

Mich.

Penn.

N.J.

eb.

Iowa

Ohio

Del.

III.

Ind.

Md.

flecked with buff or cinnamon brown, while the males have uniformly black tails or black tails flecked with gray.

Immatures resemble adults of their sex, but the two outer juvenal primaries are more pointed than the others and (at least in western birds) are narrowly marked with buff rather than whitish coloration on the outer webs. It has also been reported that the tip of the ninth primary in immature Alaskan spruce grouse is mottled and edged with brown, while in adults it is only narrowly edged with brown.

Habitat and foods. The overall geographic distribution of the spruce grouse generally conforms to that of the northern coniferous forest of Canada and the coniferous forest of the Rocky and Cascade mountains. In the eastern portion of its range the species is primarily associated with jack pine, balsam fir, and black spruce. Farther west, lodgepole pine is commonly associated with the spruce grouse's habitat. In most areas the needles of the available species of pine, spruce, and fir are major foods from late fall through spring, supplemented by berries and leaves of blueberries and other berry-producing shrubs. In general, the birds prefer rather open forests which have a considerable development of ground cover and smaller trees and shrubs to mature, dense stands of conifers. Chicks initially eat the usual assortment of insects, then various berries, and later needles. Berries are also an important late-summer food for adults of both sexes.

Social behavior. In early spring males establish display territories in moderately dense stands of coniferous or mixed coniferous-deciduous forest, where they display both from tree perches and on the ground. Openings in the forest are utilized for display flights, which vary from short, nearly vertical flutter flights to longer flights to the edge of a clearing. At least in the western (Franklin) race, the birds produce a double clapping sound during this flight by striking the wings together twice above the back. The basic territorial advertisement display of males on the ground is strutting, in which the bird stands with tail cocked, neck erect, wings slightly drooped, and the crimson eye-combs engorged. Although the neck feathers are raised in a distinctive manner that emphasizes their color pattern, no bare skin is exposed. The major sound that is produced is a tail-rustling noise, although hissing may be uttered when a female is near, and an extremely soft hooting note has sometimes also been reported. The more intensive form of strutting that is performed in the presence of a female or another male includes tail-flicking and head-jerking movements. Like the blue grouse, male spruce grouse evidently do not assist in nesting or brood-rearing activities, although males may often be seen in the presence of females with older broods, apparently because of sexual attraction.

Reproductive biology. The female usually places her nest in a well-concealed location, often under low branches, in brush, or in deep mosses in or near spruce thickets. Early estimates of clutch sizes were far too large, and most current figures suggest an average clutch of only 6 or 7 eggs. Likewise, the incubation period is now known to be 23 days, or not very different from that of the blue grouse. Until the chicks begin to fledge at about 10 days of age, the female is highly protective of them and when alarmed may make threatening movements similar to the male's strutting displays. As the chicks become older, the female, when threatened, tends to utter warning calls but not to attack the intruder, or performs distraction displays to lure the intruder from the brood. Females with older broods are often joined by males, and the birds gradually move into their winter habitat.

Willow Ptarmigan
Lagopus lagopus (Linnaeus) 1758

Other vernacular names. Arctic grouse, red grouse (Scotland form), Scottish grouse, white grouse, willow grouse, willow partridge.

Range. Circumpolar. In North America from northern Alaska, Banks Island, Melville Island, Victoria Island, Boothia Peninsula, Southampton Island, Baffin Island, and central Greenland south to the Alaska Peninsula, southeastern Alaska, central British Columbia, Alberta, Saskatchewan, Manitoba, central Ontario, central Quebec, and Newfoundland.

Identification. Adults, 14–17 inches long. All ptarmigan differ from other grouse in that (except during molt) their feet are feathered to the tips of their toes (winter) or base of their toes (midsummer), and their upper tail coverts extend to the tips of their tails. The primaries and secondaries of all the North American populations of this species are white in adults throughout the year. Males have a scarlet "comb" above the eyes which is most conspicuous in spring, and during spring and summer are extensively rusty hazel to chestnut with darker barring above except for the wings and tail. The tail feathers are dark brown, tipped with white, except for the central pair, which resemble the white upper tail coverts. In summer females lack the male's chestnut color and are heavily barred with dark brown and ocher. In autumn the male is considerably lighter, and the upper parts are heavily barred with dark brown and yellowish markings, lacking the fine vermiculated pattern found in males of the other ptarmigans at this season. The female in autumn resembles the male but is more grayish above and more extensively white below. In winter both sexes are entirely white except for the tail feathers, of which all but the central pair are dark brownish black. These may be concealed by the long coverts. In addition, the shafts of the primaries are typically dusky and the crown feathers of males are blackish at their bases. In first-winter males and females the bases of these feathers are grayish.

Field marks. The dark tail of both sexes at all seasons distinguishes the willow ptarmigan from the white-tailed ptarmigan but not from the rock ptarmigan. In spring and summer the male willow ptarmigan is much more reddish than the rock ptarmigan, and although the females of these species are very similar, the willow ptarmigan's bill is distinctly larger and higher and is grayish at the base. In fall males are more heavily barred than male rock ptarmigan, and females likewise have stronger markings than female rock ptarmigan. In winter males lack the black eye markings of the male rock ptarmigan; but since this mark is usually lacking in rock ptarmigan females, the heavier bill should be relied upon to distinguish female willow ptarmigan.

Age and sex criteria. *Females* lack the conspicuous bright reddish "eyebrows" of adult males, are more grayish brown and more heavily barred on the breast and flanks than males, and lack the distinctive rusty brown color of males in summer. In fall, females are somewhat grayer above and more heavily barred on

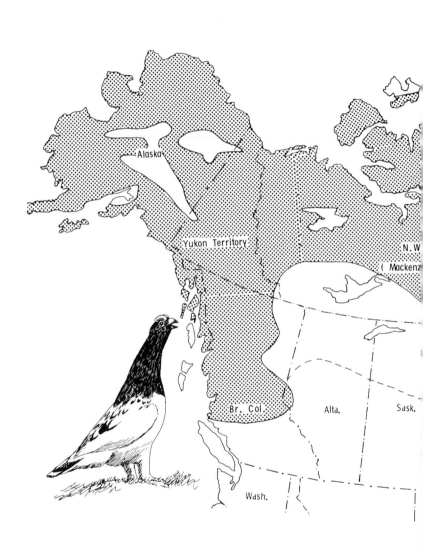

FIGURE 6. Current North American distribution of the
willow ptarmigan. Dashed line indicates normal southern
wintering limits.

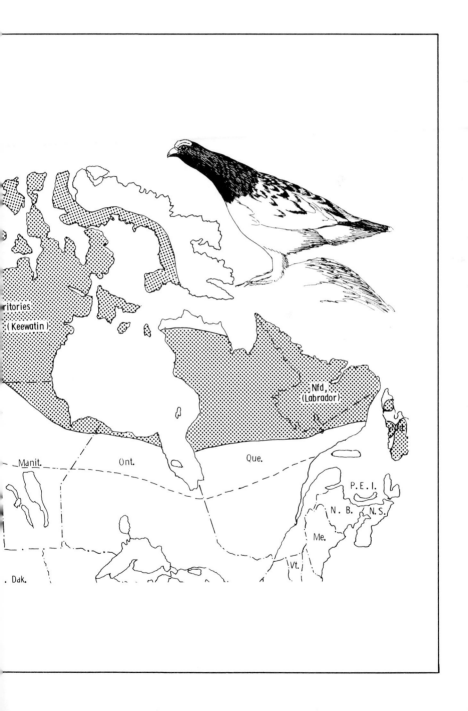

the breast and flanks than males. In winter they resemble males but the concealed bases of the crown feathers are more grayish.

In *immatures* in first-winter plumage, the tip of the tenth primary tends to be more pointed than the inner ones, but a more reliable criterion is the fact that (1) there is little or no difference in the amount of gloss on the three outer primaries of adults, whereas immatures have less gloss on the outer two primaries than on the eighth, and (2) there is about the same amount of black pigment on primaries eight and nine (sometimes more on eight than on nine) of adults whereas juveniles have more on the ninth than on the eighth.

Habitat and foods. As its name implies, the willow ptarmigan is closely associated with willow for much of the year. It usually spends winter in willow thickets along streams or near timber line, making the buds and twigs of willows a substantial part of its winter diet. In spring, males establish territories in more open areas which have elevations like rocks or hummocks that serve as territorial display sites. Nesting usually occurs in shrubby vegetation where the nest is well screened, and the chicks remain in areas that have shrub thickets for escape cover, where they feed on insects and plant materials. During the spring and summer the adults eat less willow buds and twigs as leafy materials, berries, and seeds become available. When the berries are exhausted and the leaves drop from low bushes, the birds return once again to willows, birch, and alder for their sustenance.

Social behavior. As winter wanes and the ptarmigan gradually return to their breeding areas, at sometimes substantial distances, the males spread out over the breeding area and become highly territorial. Males that become territorially established spend much of their time on exposed lookout perches, threatening or attacking intruders with displays that include erection of the red eye-combs, loud calling, and territorial "song-flights." Females are attracted to the more vigorous males by this behavior, and the male performs a number of courtship postures when confronted with a female. Both sexes have a wide variety of common vocal signals, but those of the females are higher in pitch. Males not only vigorously defend their territory against intruders, but usually also are actively involved in nest and brood defense. Willow ptarmigan are normally monogamous.

Reproductive biology. The female builds her nest within her mate's territory, usually on hard, well-drained ground, in shrubby cover close to sources of grit, water, and suitable foraging areas for the chicks. In North America the normal clutch size ranges from 7 or 8 eggs in Alaska to as many as 10 or 11 in Newfoundland. The eggs are laid on a nearly daily schedule, and the incubation period is 21–22 days. Although incubation is the female's responsibility, the male normally remains with the family through the brood-rearing period. The chicks begin to make short flights when a week or so of age, and by the age of 10–12 weeks the males' voices begin to differ from those of females. Immature birds and females begin their movement to winter cover before adult males and usually move considerably farther from the breeding areas.

Rock Ptarmigan
Lagopus mutus (Montin) 1776

Other vernacular names. Arctic grouse, barren-ground bird, rocker (in Newfoundland), snow grouse, white grouse.

Range. Circumpolar. In North America from northern Alaska, northwestern Mackenzie, Melville Island, northern Ellesmere Island, and northern Greenland south to the Aleutian Islands, Kodiak Island, southwestern and central British Columbia, southern Mackenzie, Keewatin, northern Quebec, southern Labrador, and Newfoundland.

Identification. Adults, 12.8–15.5 inches long. Both sexes carry blackish tails throughout the year. The male's scarlet comb is most evident during the spring but is apparent to some extent through the summer. In the summer males are extensively but rather finely marked with brownish black and various shades of brown, and lack the rich chestnut tone of male willow ptarmigan. In summer females are more coarsely barred and generally lighter overall than males, with somewhat finer markings than female willow ptarmigan have. Their throats and breasts are definitely barred rather than finely barred or vermiculated as in males. In autumn males are generally pale above, with tones of ashy gray predominating (or tawny brown in some Aleutian races). Females at this time have more brown and fewer black markings, plus a sprinkling of white winter feathers. Both sexes in winter are mostly white with blackish tails, and males (and some females) have a black streak connecting the bill with the eye and extending somewhat behind the eye.

Field marks. The smaller, weaker, and entirely black bill of the rock ptarmigan is sometimes detectable in the field and serves to distinguish this species from the willow ptarmigan in all seasons. In the winter, the presence of a black line through the eyes may also identify the rock ptarmigan, but its absence does not exclude this species. For plumage distinctions useful in separating willow and rock ptarmigans, see the preceding account of willow ptarmigan. During the breeding season the rock ptarmigan is found in higher, rockier, and drier country than the willow ptarmigan, but they may occur together during winter and intermediate periods. In all seasons the dark tail distinguishes the rock ptarmigan from the white-tailed ptarmigan.

Age and sex criteria. *Females* lack the reddish "eyebrows" of adult males and in summer are more heavily barred with dark markings both above and below. In autumn the female is more lightly barred but is still somewhat more heavily marked than the grayish and finely vermiculated male. In winter the sexes are nearly identical, but females usually lack the black stripe through the eye that is present in males.

Immature females in autumn are browner and more narrowly barred with blackish brown above and on the breast than are adult females. The pointed condition of the outer primaries has been reported to be an unreliable indicator of age. Instead, young rock ptarmigan may be distinguished by the fact that in adults the ninth primary (second from outside) has the same amount of

FIGURE 7. Current North American distribution of the rock ptarmigan. Dashed line indicates normal southern wintering limits.

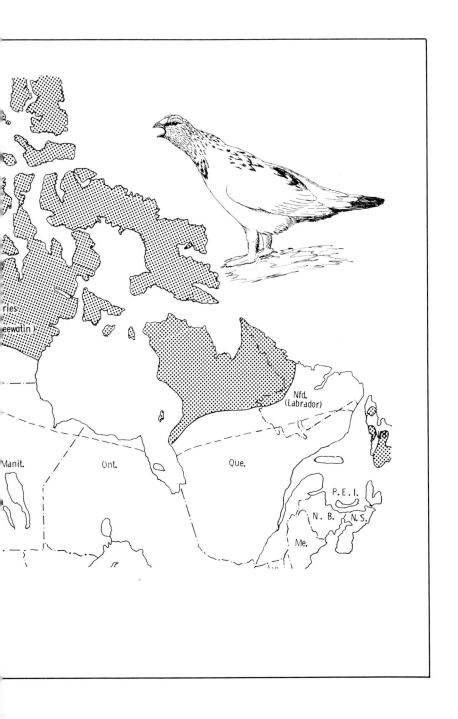

pigment as the eighth, or less, whereas immature birds have more pigment on the ninth.

Habitat and foods. The rock ptarmigan is somewhat more arctic-adapted than the willow ptarmigan, and thus is more widely distributed through the high arctic of North America and Eurasia. Although rock ptarmigans may migrate substantially south of their breeding grounds in winter, the birds often winter on shrubby slopes at timber line, in large forest openings, or even in willow thickets along rivers. Usually little winter contact between rock and willow ptarmigan occurs, and by early spring the males move into areas of more open and often rocky vegetation than is used by the latter species. Nesting habitat is typically dry, rocky, and rather barren, while brooding habitat includes swales on ridges and upper slopes, but not necessarily heavy shrub cover. Fall, winter and early spring foods are mostly the buds, catkins and twigs of birches and willows, while summer foods include a wide array of leaves, flowers, berries, and seeds. Berries such as blueberries and crowberries are important fall foods for both adults and juveniles, and as these become less available the birds return to their winter diet of birch and willow.

Social behavior. The breakup of winter flocks and the establishment of territories by males probably coincide with the emergence of breeding sites from snow cover. Males may establish territories only a few days after their return to the breeding grounds, and single birds often defend surprisingly large territories of up to a square mile. From his various lookouts the resident male calls, engages in song-flight displays, and attacks intruding males. Females are attracted to these males, and after a short period of intensive display by males, they gradually come to associate themselves with a specific male and his territory. Sometimes 2 or even 3 females will mate with a single male and nest on his territory. To what extent the male defends the female and her nest is still controversial, although a few instances of brood defense by males have been reported.

Reproductive biology. Rock ptarmigan females build simple, shallow nests, the depressions often being little more than might be caused by the weight and movements of the brooding hen. The eggs are laid at an approximately daily rate. North American records of nests indicate an average clutch of about 7 eggs, with considerable yearly variability. The incubation period is 21 days, and after hatching the female is highly attentive of her young. Females with broods apparently do not always remain within the limits of the male's territory, but instead tend to congregate on higher, moist and gently sloping areas where a variety of grasses, shrubs, and herbaceous broad-leaved plants occur. As the broods mature they tend to merge, and these flocks in turn attract groups of males and nonproductive females. Thus flocks of several hundred may build up during the fall before the birds begin their movement to winter cover.

White-tailed Ptarmigan

Lagopus leucurus (Richardson) 1831

Other vernacular names. Snow grouse, snow partridge.

Range. From central Alaska, northern Yukon, and southwestern Mackenzie south to the Kenai Peninsula, Vancouver Island, the Cascade Mountains of Washington, and along the Rocky Mountains from British Columbia and Alberta south to northern New Mexico.

Identification. Adults, 12–13.5 inches long. In any nonjuvenal plumage the white tail will serve to distinguish this species from the other two ptarmigans. Adult males in summer plumage are vermiculated and barred or mottled with black, buff, and white on the back, with a buff or pale yellowish tone predominating on the lower back, rump, and upper tail coverts. The underparts are mostly white. Unlike the other ptarmigans, white-tailed have completely white wings and tail (except for the central pair of feathers) at this season. Females are similar in plumage but have a

FIGURE 8. Current distribution of the white-tailed ptarmigan.

heavily spotted and more yellowish color on the back. In the fall both sexes are mostly pale cinnamon or reddish brown above, with fine spotting and vermiculations of brownish black, and with a lighter head and neck. A few breast feathers are usually marked with white, and the abdomen, under tail coverts, tail, and wings are white. In the winter both sexes are pure white except for a black bill, eyes, and claws.

Field marks. This species is a small alpine ptarmigan with white wings and tail in summer and entirely white plumage in winter. It is usually extremely difficult to see against a lichen-covered rocky background and is therefore often overlooked unless forced to fly.

Age and sex criteria. *Females* (unlike those of the two other ptarmigan species) exhibit eye-combs virtually identical to the adult male's. In summer, however, hens are more coarsely and regularly barred with black and rich yellowish buff markings on their brownish back and side feathers, while the male's feathers in these areas are finely vermiculated with brown and black. In addition, although males retain their white lower breast, abdomen, and under tail coverts through the summer, females have yellowish buffy brown feathers with some black barring in these areas. In the autumn differences between the sexes diminish, but for a time females retain a few of their coarsely barred nuptial plumage feathers, especially on the nape, sides, inner wing, and upper tail coverts. In winter the sexes are identical in plumage but may differ slightly in wing length and length of the outer five primaries and the outer tail feathers, which average longer in males. In spring, males can be recognized by their distinctive black-tipped head and neck feathers, which provide a hooded effect that is lacking in females as they gradually acquire their brown, black, and yellow nuptial plumage.

Immatures may be recognized by the pigmentation of their two outer primaries and the outer primary covert. If black pigment occurs on either the ninth or tenth primary and the outer primary covert, the bird may confidently be called an immature; adults lack pigmentation in these areas.

Habitat and foods. The wintering habitat of this species apparently must contain alpine willows. Willows are also essential for the establishment of male territories. A suitable summer habitat for nesting birds includes adequate brooding areas, primarily alpine meadows with short vegetation consisting mainly of sedges and herbaceous broad-leaved plants. During the summer the leaves and flowers, seeds, and bulbils of these plants are important foods for adults and for chicks more than a few weeks old. By late summer willow becomes an important component of the diet, and willow buds and twigs are practically the sole source of winter food for Colorado ptarmigan. In Alaska the birds evidently rely to a much greater extent on alder than on willows, perhaps because of the greater availability of alder in these more northerly areas.

Social behavior. With the return of males from their timberline wintering areas to the alpine breeding grounds, they gradually establish territories, often those they defended in previous years. These territories range from about 15 to nearly 50 acres in size. The strength of territorial defense increases greatly when the females arrive on the breeding grounds. Males seek to attract them, and to repel other males, by an aerial display, or "scream flight," "ground challenging," and various intimidation displays. The scream flight is equivalent to the song flight of the other ptarmigan species, and the ground challenging consists of screaming from various calling posts in the territory. When unmated females enter a male's territory, he performs a "courtship chase" resembling aggressive attacks toward other males. Evidently, pair formation is achieved by the repeated performance of this and other displays. White-tailed ptarmigan are normally monogamous, although some males succeed in attracting 2 females. The pair bond may last only 2 or 3 weeks, and usually is terminated by the time of hatching.

Reproductive biology. Females build their nests in a variety of locations; their strongly protective coloration makes it difficult for predators to see them. The clutch size is usually small, from 4 to 7 eggs, which are laid at intervals of about 1½ days. The incubation period is from 22 to 23 days. Females typically defend their nests and broods very strongly, performing distraction displays while sitting on eggs or tending very young chicks. As they become older, she tends to place herself between the intruder and the brood, running back and forth while hissing. Males have at times been reported to defend the nest site, but not the broods. Fledging occurs when the chicks are only about 10 days old, and the females and broods soon move to areas that provide a combination of rocky habitat and an abundance of short herbaceous vegetation. Hens remain with their well-grown broods through autumn, as the birds gradually move toward their lower-altitude wintering areas.

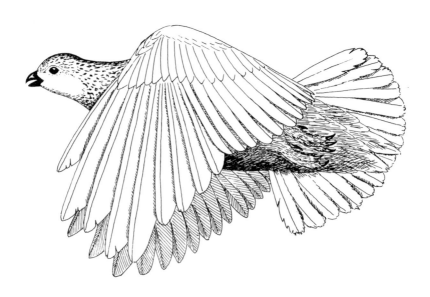

Ruffed Grouse
Bonasa umbellus (Linnaeus) 1776

Other vernacular names. Birch partridge, drummer, drumming grouse, long-tailed grouse, mountain pheasant, partridge, pine hen, pheasant, tippet, white-flesher, willow grouse, wood grouse, woods pheasant.

Range. Resident in forested areas from central Alaska, central Yukon, southern Mackenzie, central Saskatchewan, central Manitoba, northern Ontario, southern Quebec, southern Labrador, New Brunswick, and Nova Scotia south to northern California, northeastern Oregon, central Idaho, central Utah, western Wyoming, western South Dakota, northern North Dakota, Minnesota, central Arkansas, Tennessee, northern Georgia, western South Carolina, western North Carolina, northeastern Virginia, and western Maryland. Recently introduced into Nevada and Newfoundland.

Identification. Adults, 16–19 inches long. Both sexes have relatively long, slightly rounded tails that are extensively barred

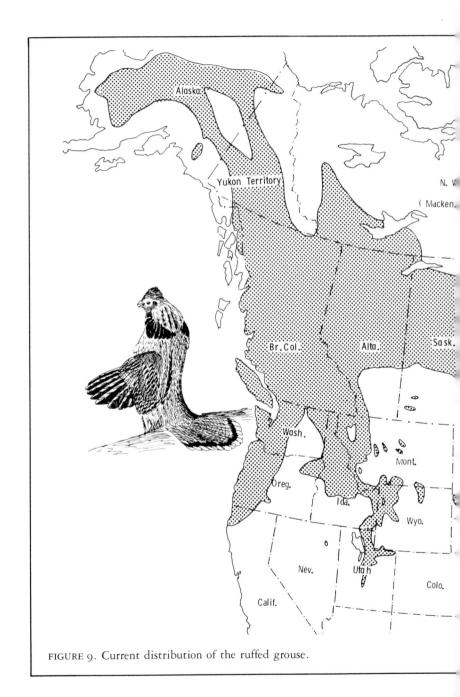

FIGURE 9. Current distribution of the ruffed grouse.

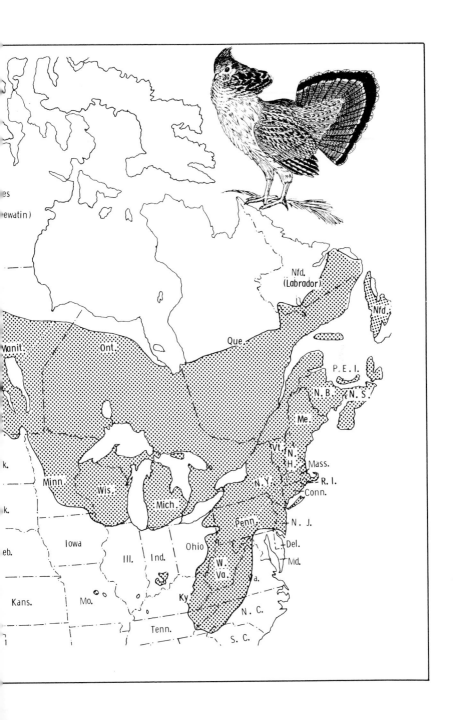

above and have a conspicuous dark band just below the tip. The neck lacks large areas of bare skin, but both sexes have dark ruffs. The feathering on the legs does not reach the base of the toes; the lower half of the leg is essentially nude. Both sexes are definitely crested, but the crest feathers are not distinctively colored. In addition, males have a small comb above the eyes that is orange red and most evident in spring. Most races exist in both gray and brown tail-color phases, which appear with the first-winter plumage. Otherwise, little seasonal, sexual, or age variation occurs. The birds are generally wood brown above, with blackish ruffs (less conspicuous in females). The tails of both sexes have 7–9 alternating narrow bands of black, brown (or gray), and buff, followed by a wider blackish band just below the tip that is bordered on both sides with gray. This band may be broken at the center in females and some (presumably first-year) males. In winter, both sexes develop horny fringes on the sides of their toes which are more conspicuous than in most other species of grouse.

Field marks. The fan-shaped and distinctively banded tail and neck ruffs of both sexes make field identification easy. The birds usually take off with a conspicuous whirring of wings, and in spring males are much more often heard drumming than they are seen.

Age and sex criteria. *Females* have shorter tails than do males and the broad black tail band is broken on their central tail feathers. Either sex may have a mottled pattern on the central tail feathers (which occurs in about 15 percent of the population), but a bird with this marking is twice as likely to be a male as a female. Females have little or no color on the bare skin over the eye, whereas in males this area is orange to reddish orange.

Immatures can be identified by the pointed condition of their two outer primaries, especially the outermost one. One researcher states that during the hunting season the condition of the tenth primary was useful for determining the age of nearly 60 percent of the birds, with only a 2 percent error. The presence of sheathing (frayed remnants of the "pin" covering of the feather) at the base of the outer two primaries (in adults) or on the eighth but not the ninth or tenth primary (in immatures) separated 79 percent of the birds examined with a 3 percent error. Immature males can also be distinguished from adults by their shorter central tail feathers, and it is reported that the two outer primaries of immatures have outer webs that are pale grayish brown and mottled or stippled with lighter buff, instead of being buff or whitish with darker brown markings.

Habitat and foods. The winter habitat of the ruffed grouse must include roosting sites and a reliable source of food, usually the buds and twigs of such trees as poplars, apples, birches, oaks, and cherries. If not snow-covered, grape vines, greenbrier, laurel, blueberry bushes, wintergreen, and similar shrubs and vines are also used for winter food and cover. With spring, males move into areas that have suitable drumming logs and females seek out a suitable nesting habitat, typically a middle-aged stand of hardwoods or mixed hardwoods and conifers. Brushy habitats like overgrown farmlands and sites of spot lumbering are favored for brood rearing. The ruffed grouse's diet varies appreciably with the locality, but in most areas aspen and poplar are important sources of food, especially in winter. The buds, twigs, and catkins of these species as well as of birches are eaten in most areas, and the use of these and other trees and shrubs may continue well into spring. In early summer, as berries and fruits become available, the diet of ruffed grouse changes markedly, and until late fall a great variety of fruits, berries, and seeds are eaten. Some of these berries or fruits may remain available into winter, supplementing the basic winter diet of buds and twigs.

Social behavior. With the onset of spring, male ruffed grouse seek out suitable territories, which are usually marked by the occurrence of well-grown (40–50-year-old) aspens and one or more drumming stages, most frequently a log. Although a territorial male may use more than one drumming log, he typically favors one, and on this he performs his characteristic wing-beating display each morning and evening. With his tail braced against the log and his claws imbedded in the partially rotted wood, he begins a series of strong wingstrokes. These strokes produce a low drumming noise that has surprising carrying power and provides an effective advertisement of his territory. When a drumming male attracts a female, he quickly assumes a strutting attitude, with tail strongly fanned, neck ruffs raised, and wings drooping. He also shakes his head in a rotary motion, hisses, and makes a short rush toward the female. Females are receptive to displaying males for only a few days, and after fertilization occurs they probably leave the male and begin their nesting activities.

Reproductive biology. After the female leaves the male, she seeks out a nest site, usually close to a clump of male aspen trees, the catkins of which she relies upon for food during incubation. Tree bases are a favorite location for nests, and logs, bushes, or brush piles are also frequently used. The clutch normally numbers about 10–12 eggs, laid at the rate of 2 eggs every 3 days. The incubation period is 24–25 days; and in contrast to most grouse, this species frequently attempts renesting if the first clutch should fail. The female normally feigns injury if disturbed during incubation, but after hatching she more often stands her ground, spreading her tail and assuming a posture much like the male's strutting display as she hisses or utters squealing notes. After the chicks are able to fly at the age of 10–12 days the normal response for the family is to flee from danger. When the young are about 12 weeks of age the families begin to break up and a general fall dispersal of the juvenile birds begins.

Pinnated Grouse
Tympanuchus cupido (Linnaeus) 1758

Other vernacular names. Prairie chicken, prairie cock, prairie grouse, prairie hen.

Range. Greater prairie chicken: current resident of remnant prairie areas of Michigan, Wisconsin, and Illinois and from southern Manitoba south to western Missouri and Oklahoma. Attwater prairie chicken: portions of the coastal plain of Texas. Lesser prairie chicken: from southeastern Colorado and adjacent Kansas south to eastern New Mexico and northwestern Texas.

Identification (greater and Attwater prairie chickens). Adults, 16–18.8 inches long. Both sexes are nearly identical in plumage. The tail is short, somewhat rounded, and the longer under (but not upper) tail coverts extend to its tip. The neck of both sexes has elongated pinnae made up of about ten feathers of graduated length that are much longer in males than in females. Males have conspicuous yellow combs above the eyes and bare

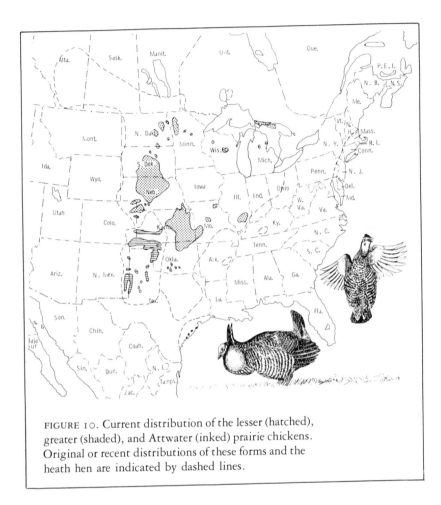

FIGURE 10. Current distribution of the lesser (hatched), greater (shaded), and Attwater (inked) prairie chickens. Original or recent distributions of these forms and the heath hen are indicated by dashed lines.

areas of yellowish skin below the pinnae that are exposed and expanded during sexual display. The upper parts are extensively barred with brown, buff, and blackish coloration, while the under parts are more extensively buff on the abdomen and whitish under the tail. Crosswise barring of the feathers is much more definite in this species than in the sharp-tailed grouse, which has darker, V-shaped markings and more white exposed on the under parts.

Identification (lesser prairie chicken). Adults, 15–16 inches long. In general resembles the greater prairie chicken, but has brown bars (with black forming narrow margins) on the back and rump instead of the darker, blackish ones typical of the greater prairie chicken. The breast feathers are more extensively barred with brown and white, and the flank feathers are barred with brown and blackish instead of only brown. Males have reddish rather than yellowish skin in the area of the gular sacs and during display their yellow combs are more conspicuously enlarged than those of greater prairie chickens. As in that form, females have shorter pinnae and are more extensively barred on the tail.

Field marks. The only species easily confused with either the greater or lesser prairie chicken is the sharp-tailed grouse, which often occurs in the same areas as greater prairie chickens. Sharp-tailed grouse can readily be recognized by their pointed tails, which, except for the central pair of feathers, are buffy white, and by their whiter under parts as well as a more "frosty" upper plumage pattern, which results from white spotting where the pinnated grouse has buffy.

Age and sex criteria (greater and Attwater prairie chickens). *Females* may readily be recognized by their shorter pinnae (those of the female greater prairie chicken have a maximum length of 1¾ inches, or 44 mm.; those of males are a minimum of 2⅜ inches, or 63 mm. long) and their extensively barred outer (rather than only central) tail feathers. The central crown feathers of females are marked with alternating buff and darker crossbars, whereas males have dark crown feathers with only a narrow buff edging. In the Attwater prairie chicken the pinnae of females are reported to be about ⁹/₁₆ inch (14 mm.) long, while those of males are over 2 inches (53 mm.).

Immatures may be recognized by the pointed, faded, and frayed condition of the outer 2 pairs of primaries (see sharp-tailed grouse account, following) The pinnae length of first-autumn males is not correlated with age.

Age and sex criteria (lesser prairie chicken). *Females* may be identified by their lack of a comb over the eyes and their brown-barred under tail coverts (in males these are black with a white "eye" near the tip). Males have blackish tails, with only the central feathers mottled or barred, while the tails of females are extensively barred.

Immatures can usually be identified by the pointed condition of the 2 outer pairs of primaries. The outermost primary of young birds is spotted to its tip, while that of adults is spotted only to within an inch or so of the tip. In addition, the upper covert of the outer primary is white at the outer end of the shaft, whereas in adults the shafts of these feathers are entirely dark.

Habitat and foods. The three populations of pinnated grouse occupy quite different habitats and depend on different sources of food. Throughout the year the greater prairie chicken is now closely associated with stands of natural prairie, especially where these natural grasslands are interspersed with moderate amounts of small grain cropland. The lesser prairie chicken has long been associated with the arid natural grasslands of the southern Great Plains, although it too presently tends to rely on cultivated grains like sorghum during winter. The Gulf Coast population of Attwater prairie chickens favors lightly grazed natural grasslands as a winter habitat. Open areas of grassland, either in elevated or, more rarely, level or depressed situations, are favored by all three forms for display grounds. Nesting cover for all of the forms is typically in ungrazed meadows, natural prairie stands, or clumps of prairie grasses. Well-drained sites that provide good concealment from above are especially favored. Brood-rearing habitats usually have heavier cover than the nesting area, with shade, a supply of insects, and succulent plants available. After the young have passed through their early insect-eating stage, they rely increasingly on plant material, especially the seeds of wild plants and grains such as corn, sorghum, oats, and wheat. Evidently adult lesser prairie chickens eat a substantially greater number of insects than do the larger races.

Greater Prairie Chicken (Male "Flutter-jump")

Social behavior. All three forms of prairie chickens are found in mixed-sex flocks during the late fall and winter, but by early spring the males return to their traditional display grounds, where they reestablish old territories or, in the case of young birds, try to acquire new ones. Like the sage grouse and sharp-tailed grouse, experienced and older male pinnated grouse tend to hold central territories, while the younger males establish peripheral ones. Territorial advertisement consists of the well-known "booming" display (called "gobbling" in the lesser prairie chicken). The male inflates the bare yellow to orange skin areas ("air sacs") on the sides of his neck, erects the feathered pinnae above his head, drops his wings, stamps his feet, and calls. Besides the rather low-pitched call, tail-clicking sounds are produced and the foot stamping is often audible. Males often also perform short vertical flights, called "flutter-jumps," usually while uttering cackling calls. In the immediate presence of females, males of all three forms sometimes perform a "nuptial bow," with wings spread, pinnae erect, and bill lowered to the ground. This display often (but not invariably) serves as a precopulatory signal. Females visit the display grounds only when receptive for copulation, and after a successful mating retire to begin nesting.

Reproductive biology. Females build their nests at varying distances from the display grounds at which they were fertilized, and may actually nest nearer to some other display ground. In all of the races the eggs are laid at an approximately daily rate, and in all the normal clutch size of initial nesting attempts is from about 12 to 14 eggs. The incubation period is from 23 to 26 days in all. The nest is usually deserted within 24 hours after the last chick has hatched, and females with newly hatched broods perform intensive decoying behavior by feigning injury. Within about 10 days the young are able to fly short distances, and thereafter both the hen and brood typically flush when disturbed. The broods usually remain with their mothers for 6–8 weeks, after which the families gradually disintegrate. A reorganization of old and young birds into fall flocks then occurs, and there is a gradual movement to suitable winter cover.

Chachalaca

Wild Turkey

Wild Turkey

Wild Turkey

Sage Grouse

Sage Grouse

Dusky Blue Grouse

Dusky Blue Grouse

Dusky Blue Grouse

Canada Spruce Grouse

Canada Spruce Grouse

Canada Spruce Grouse

Willow Ptarmigan

Willow Ptarmigan

Willow Ptarmigan

Rock Ptarmigan

Rock Ptarmigan

White-tailed Ptarmigan

White-tailed Ptarmigan

White-tailed Ptarmigan

Ruffed Grouse

Ruffed Grouse

Sharp-tailed Grouse

Sharp-tailed Grouse

Greater Prairie Chicken

Greater Prairie Chicken

Lesser Prairie Chicken

Lesser Prairie Chicken

Sharp-tailed Grouse

Pedioecetes phasianellus (Linnaeus) 1858

Other vernacular names. Brush grouse, pintail grouse, prairie pheasant, sharptail, speckle-belly, spike-tail, spring-tail, white-belly, white-breasted grouse.

Range. Currently from north-central Alaska, Yukon, northern Mackenzie, northern Manitoba, northern Ontario, and central Quebec south to eastern Washington, extreme eastern Oregon, Idaho, northeastern Utah, Wyoming, and Colorado, and in the Great Plains from eastern Colorado and eastern Wyoming across Nebraska, the Dakotas, northern Minnesota, northern Wisconsin, and northern Michigan.

Identification. Adults, 16.4–18.5 inches long. The sexes are nearly identical in plumage. The tail is strongly graduated in both sexes with the central pair of feathers extending far beyond the others, but the tips are not pointed. Both sexes are feathered to the base of the toes, and males have an inconspicuous yellow comb (somewhat enlarged during display) and paired pinkish to

FIGURE 11. Current (shaded) and recent (dashed) distributions of the sharp-tailed grouse.

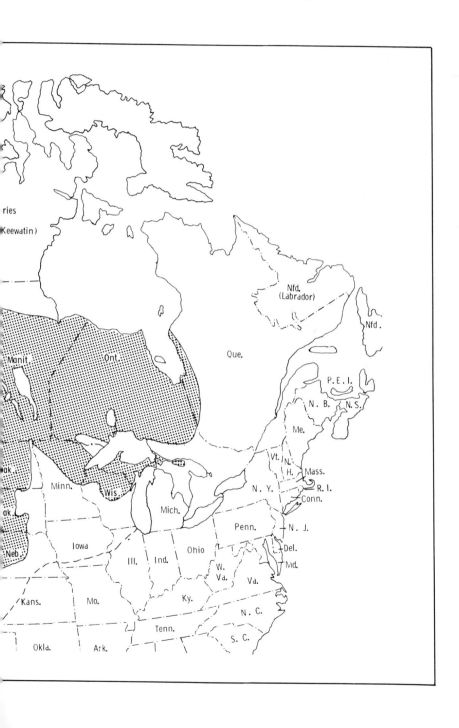

pale violet areas of bare neck skin that are also expanded during display, though not to the degree found in prairie chickens. Both sexes have inconspicuous crests, and the head and upper parts are extensively marked with barring and spotting of white, buffy, tawny brown, and blackish coloration. White spotting is conspicuous on the wings, and the relative amount of white increases toward the breast and abdomen, which are pure white. The middle pair of tail feathers is elaborately patterned with brown and black, but the others are mostly white. The breast and flanks are intricately marked with V-shaped brown markings on a white or buff background.

Field marks. The habitat of this species varies considerably throughout its range from grassland or edge to scrub forest, but the bird is basically to be found in fairly open country, where its pale, mottled plumage blends well with its surroundings. In flight the white under parts are conspicuous, as is the whitish and elongated tail. On the ground, the birds have a much more "frosty" appearance than do prairie chickens, which are generally darker and lack definite white spotting.

Age and sex criteria. *Females* may be identified with about 90 percent reliability by a crosswise barred pattern on the central tail feathers (males have more linear markings). Also, the crown feathers of females have alternating buff and dark brown crossbars, whereas the male's crown feathers are dark with buff edging.

Immatures may be identified by the usual characteristic of pointed outer primaries, but one researcher suggests that a comparison of the eighth and ninth primaries for relative amounts of wear (equal or little wear on either in adults, greater wear on the ninth in immatures) is the most suitable method of judging age in prairie grouse.

Habitat and foods. Although generally considered a plains grassland species, the sharp-tailed grouse in some parts of its range is also associated with sagebrush semidesert, open woodlands, and brushy coniferous or deciduous forests. A combination of wooded and nonwooded winter habitats provides opportunities for browsing in the trees and also a possible source of grain or similar foods. Spring habitat needs include suitable areas for display grounds, which are generally sparsely vegetated and somewhat elevated sites, although marshes serve this purpose in a few areas. Nesting habitats are somewhat more diverse than those used by pinnated grouse, but typically are well concealed from above and are usually in natural rather than cultivated vegetation. As the broods hatch they move to somewhat heavier woody cover with an abundance of herbaceous plants and shrubs in forest openings. The sharp-tail reportedly eats insects to a lesser extent than does the pinnated grouse during summer, preferring succulent plants like dandelions and buttercups. From late fall through spring the diet of the two species is nearly identical where they occur together.

Social behavior. The general social organization of sharp-tailed grouse and pinnated grouse is nearly identical, the only major difference being the nature of the male displays performed on the display ground, which in this species is called a "dancing ground." This term reflects the most conspicuous aspect of the male's territorial advertisement display. Instead of foot stamping in place, the male moves forward in a circular pattern with his wings rigidly outstretched, his tail cocked and shaken vigorously in rhythm with his stepping movements. The noise produced is primarily mechanical, as a result of the tail feathers being scraped over one another and the feet alternately striking the ground. He has several calls as well, and performs a "cooing" display that is comparable to the "booming" of the pinnated grouse although it has a somewhat different functional role. Females visit the ground prior to sunrise, and most copulations have been completed by the most dominant male, the "master cock," or his immediately neighboring territorial occupants by shortly after sunrise. After each female is fertilized she rapidly leaves the display ground, and probably does not return again unless she is forced to renest.

Reproductive biology. The female begins to construct a nest in a concealed site at about the time she is fertilized, or possibly even before. The eggs are laid on an approximately daily basis until the clutch of around 12 is complete. Incubation thereafter requires 23–24 days; and although some renesting attempts are made by unsuccessful females, they probably do not substantially contribute to the year's production of offspring. After the brood has hatched, the nest site is deserted rapidly and the brood is led to areas where insects and succulent green foods are available. By the time they are 10 days old the chicks are flying very short distances, and after 6–8 weeks they are essentially independent of their mother. At this time the broods break up and the young disperse, sometimes moving fairly great distances.

Mountain Quail
Oreortyx pictus (Douglas) 1829

Other vernacular names. Mountain partridge, painted quail, plumed quail.

Range. Resident in the western United States from southern Washington and southwestern Idaho east to Nevada and south to Baja California. Also introduced into western Washington and western British Columbia (Vancouver Island). Introduced but of doubtful status in western Colorado.

Identification. Adults, 10.6–11.5 inches long. The sexes are very similar in appearance. This relatively large western quail differs from all others in that both sexes have straight, narrow, blackish crests composed of only two feathers, which appear with the juvenal plumage. The throat is chestnut, edged with black and separated from the slate gray chest, neck, and head by a white line. Otherwise the birds are plain olive gray on the back, wings, and tail. The flanks are rich, dark brown, with conspicuous vertically oriented black and white bars.

Field marks. The slender plumes and boldly patterned flanks serve to identify mountain quail without difficulty. The California quail may occur in the same areas but has a shorter, curved crest of teardrop shape and dull brown flanks that are narrowly streaked with white. A loud, clear, whistled *quee-ark* or *plu-ark* is the advertising call of the male in spring.

Age and sex criteria. *Females* have slightly shorter and browner plumes than males (an average of 12 birds is 58 mm., or 2¼ inches, with a maximum of 66 mm., or 2⅝ inches, compared to a minimum of 66 mm. and an average of 72 mm., or 3⅛ inches, for 12 males). One researcher reports that the male is also more brightly colored beneath and the gray of the hind-neck is more sharply defined than that of the female. Another observer likewise notes that in females the brown back color extends to the top of the head, while in males the back of the neck is grayish blue. The neck color is probably the most reliable criterion for distinguishing sex but has limited use with dark coastal birds.

Immatures have buff-tipped greater primary coverts rather than the uniform gray found in adults. The two outer primaries are also more pointed and frayed than the inner ones in immatures.

Habitat and foods. The winter habitat typically consists of mixed trees, brush, and herbs, which provide a diverse array of seeds and mast. Acorns and, at times, pine seeds are major items of the winter diet. With spring, the birds move to higher altitudes offering a combination of open brush and tree cover on sloping topography where the birds can escape by running uphill. Nesting habitats usually include a source of water and succulent plants and insects, which are important foods for chicks. Fall habitats must provide abundant food, usually in the form of ripening acorns and pine seeds, plus the fruits of a large variety of shrubs and vines.

Social behavior. Throughout nearly the entire year the covey is the basis for the social structure of the American quails; and it

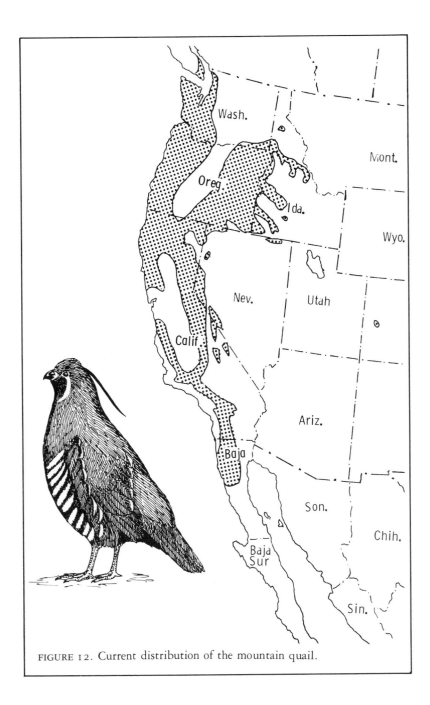

FIGURE 12. Current distribution of the mountain quail.

generally is small, from 5 to 10, or not much bigger than a single family plus a few adults that were unsuccessful in nesting. Mate selection begins in early spring, while the coveys are still intact. At that time unmated males select whistling posts, from which they utter repeated *plu-ark* calls at the rate of about 8 or 10 per minute. Rather than advertising a defended territorial area, these notes simply proclaim the presence of an unmated male, and thus serve more to attract available females than to ward off other males. Mated pairs become antagonistic toward other quails, and soon spread out over the available nesting habitat.

Reproductive biology. The nests are usually well concealed, often under fallen pine branches, amid weeds or shrubs at the base of large trees, beside large, shaded rocks, or in masses of shrubby vegetation. Clutch sizes of initial nesting attempts average about 9 or 10 eggs, and incubation lasts 24–25 days. The male takes an active role in nest and brood defense, and when disturbed will perform distraction displays toward intruders. Should the female be killed, males apparently will take over incubation, and sometimes broods are seen being tended by only a single bird of either sex. Although renesting after a nest failure is typical, there is no evidence yet to support the possibility that 2 broods are normally raised in a single season. By late summer the well-grown broods, supplemented by lone birds or unsuccessful nesters, move gradually to lower elevations.

Scaled Quail
Callipepla squamata (Vigors) 1830

Other vernacular names. Blue racer quail, blue quail, cotton-top quail, Mexican quail, scaled partridge, top-knot quail.

Range. From southern Arizona, northern New Mexico, eastern Colorado, and southwestern Kansas south to central Mexico. Introduced into central Washington and eastern Nevada.

Identification. Adults, 10–12 inches long. The sexes are very similar in plumage. Scaled quail have a predominantly bluish gray coloration (thus "blue quail"), and are extensively marked on the back, breast, and abdomen with blackish "scaly" markings. The crest is bushy, varying in color from buff in females to more whitish in males. Otherwise, the head is light grayish brown; the lower back, wings, and tail are brownish gray to gray; and the flanks are grayish to brownish with lighter markings along the shafts. Males of one race have chestnut coloration on the abdomen similar to that of male California quail.

Field marks. The "cottontop" crest is often visible from some distance, and the generally grayish coloration of this species sets it apart from all other quail in the arid habitats where it occurs. The birds are usually reluctant to fly, preferring to run rather than remain hidden. The distinctive *pey-cos* location calls (stronger in males) often reveal the presence of scaled quail in an area.

Age and sex criteria. *Females* may be distinguished from adult males by their less conspicuous crests and by the dark brown streaks on the sides of the face and the throat (males have unstreaked pearly gray to white coloration in this area).

Immatures of both sexes have buff-tipped greater primary coverts associated with the first seven primaries.

Habitat and foods. The distribution of the scaled quail largely conforms to the limits of the Chihuahuan desert and the adjacent desert grasslands. The species seems to survive best where there is a combination of annual weeds, some shrubby or spiny ground cover, and available surface water. In the northern part of its natural range it is especially associated with sandy soils and sand sagebrush vegetation. Suitable nesting cover consists of a variety of shrub and broad-leaved herb species, overhanging rocks, and even old machinery and junk. A source of midday shade and loafing cover is important during summer, but it must not be so thick as to prevent escape by running. Scaled quail eat a large variety of seeds of weeds, grasses, shrubs, and trees, showing little preference for a specific food in most areas. Likewise, the eating of insects seems to vary greatly with locality and year.

Social behavior. The fairly large winter coveys of scaled quail remain intact until the males begin to come into reproductive condition, when the combination of increasing male-to-male aggression and the separation of paired birds from the coveys causes their dissolution. At this time unmated males begin their *whock* calling from exposed sites or crowing perches. Although this

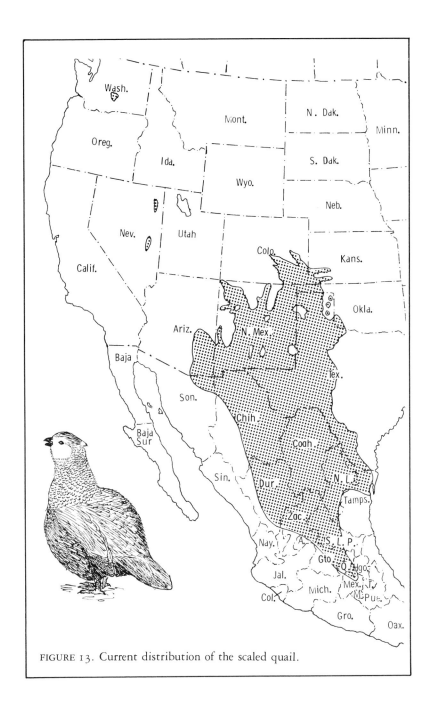

FIGURE 13. Current distribution of the scaled quail.

species is still little studied, such calls seem to function like the *bob-white* calls of that species in attracting unmated females. In spite of a long potential nesting season, actual egg laying by paired females may be deferred until the start of the summer rainy season.

Reproductive biology. Nests are usually located under shrubs or in some other protected and shaded site, and a clutch of 12 or 14 eggs is typical for initial nesting attempts. Incubation requires from 22 to 23 days; and although males remain near the nest and help protect it, there has only been a single reported instance of a male being observed incubating. However, it is known that the male sometimes takes over the care of the newly hatched brood, allowing the female to begin a second clutch. There is a high rate of nest losses from various causes, and during unusually dry years the birds may not even attempt to nest. Repeated nesting attempts usually allow for the eventual hatching of a brood, of which roughly half are likely to survive to the start of the fall hunting season. As the chicks mature, broods gradually become organized into coveys of 20–50 or more birds.

Gambel Quail

Lophortyx gambelii Gambel 1843

Other vernacular names. Arizona quail, desert quail.

Range. From southern Nevada, southern Utah, and western Colorado south to northeastern Baja California, central Sonora, northwestern Chihuahua, and western Texas.

Identification. Adults, 9.5–11 inches long. The sexes are different in appearance. This southwestern quail has a blackish, forward-tilting teardrop-shaped crest like the California quail but completely lacks the scaly patterning of the under parts typical of the latter. Only on the back of the neck of males is some scaly patterning evident, and this is ill-defined. Male Gambel quail also have a black forehead and reddish-brown crown coloration, and both sexes have more reddish brown flank coloration than the California quail. Otherwise the birds are generally grayish brown to brown on the upper parts and tail and have either buff under parts that may be streaked with brown (females) or an extensive black area on the abdomen (males). Males also have a conspicuous black throat that is lacking in females.

Field marks. Generally limited to desert regions of the southwest, Gambel quail can be identified in the field by the combination of its teardrop crest and unscaled under parts. The rich reddish-brown flanks of both sexes are visible at considerable distances, and at close range the reddish crown color of males and the black mottling of their under parts may be evident. This species' calls are similar to those of the California quail, but are less metallic and more nasal. The distinctive location call consists of occasionally repeated *chi-ca-go-go* notes (occasionally California quail will also add a fourth syllable to their location call).

Age and sex criteria. *Females* have dark brown rather than black crests and lack black throats.

Immatures have mostly buff-tipped greater primary coverts for the first year. The outer two primaries may be somewhat more pointed and frayed than the inner ones in immature birds.

Habitat and foods. The range of the Gambel quail centers on the Sonoran desert, an area that is characterized by low annual precipitation, mild winter temperatures, and generally low elevations. In the western part of its range the species also occurs in upland deserts, such as the Mohave region of Arizona, California, and Nevada, and it likewise is found up the Colorado River basin some distance into Utah. Soils and climates that produce a diversified and luxuriant shrub growth as well as herbaceous growth during the breeding season are favored by this species. Nesting and brooding habitats include shade, brushy escape cover, and foraging sites where insects and succulent green plants are plentiful. Surface water is apparently not needed where succulent plants are available. Leafy material, especially of legumes, and seeds of a wide variety of herbaceous plants are the primary adult foods, with animal materials, fruits, and mast all of minimal significance.

Social behavior. The coveys of Gambel quail normally consist of several family units, together with varying numbers of non-

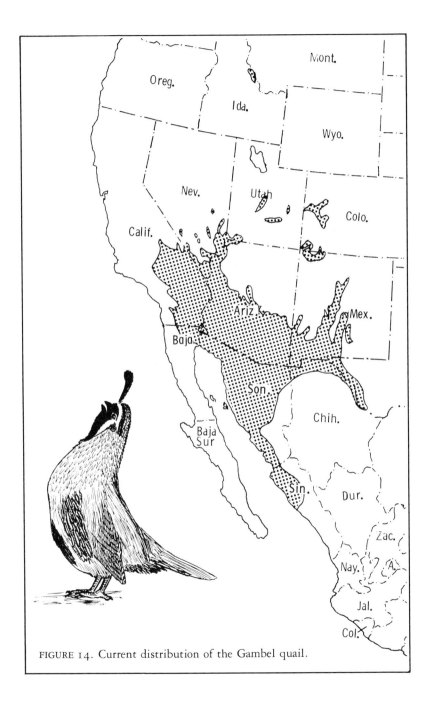

FIGURE 14. Current distribution of the Gambel quail.

breeder adults. By late winter, the coveys begin to break up as pairs separate from them and aggression increases among the unmated males. Such birds establish crowing posts, from which they repeatedly utter their distinctive *kaa* calls. Pair formation is evidently a subtle process, occurring over a prolonged period of contact, and is probably initiated by the male chasing the female in a display that differs only slightly from aggressive male-to-male chases. Once a pair bond is formed, the birds are strongly monogamous.

Reproductive biology. Females usually locate their shallow nest scrape in the shade of desert trees or shrubs. A clutch of about 12–14 eggs is laid on an egg-per-day basis, and incubation begins with the laying of the last egg. It lasts 21–23 days, and is performed by the female alone, with the male usually sitting on a lookout perch some 40–80 feet away. If the nest is approached by an intruder, the male typically performs an injury-feigning distraction display. Although the male has not been reported to undertake incubation, it is well established that he sometimes takes over the care of the newly hatched brood, allowing the female to begin a second clutch. Or, after a month or so of care by the pair, the chicks may be left with other older birds and a second clutch initiated. Like other grouse and quail chicks, they initially feed almost exclusively on insects but soon begin to eat succulent vegetation and eventually are almost totally vegetarians.

California Quail

Lophortyx californicus (Shaw) 1798

Other vernacular names. California partridge, crested quail, topknot quail, valley quail.

Range. From southern Oregon and western Nevada south to the tip of Baja California. Introduced into southern British Columbia, Washington, Idaho, northern Oregon, and Utah.

Identification. Adults, 9.5–11 inches long. The sexes are different in appearance. This widespread quail of the western foothills resembles the Gambel quail inasmuch as both sexes have forward-tilting, blackish, teardrop-shaped crests. Both sexes also have clear bluish gray to gray chests that become buff toward the abdomen and darker "scaly" markings similar to the scaled quail's. The flanks are brownish gray with lighter shaft streaks, and the upper parts are generally gray to brownish gray, intricately marked with darker scaly markings. Males have black throats and a chestnut-tinged abdomen and are chocolate brown behind the plume, while the area in front of the eyes and above the bill is whitish.

Field marks. The combination of a teardrop crest and scaly markings on the lower breast and abdomen is distinctive for both sexes. Males of this species may be distinguished from the very similar Gambel quail by the combination of a whitish rather than blackish forehead, the lack of a black abdomen patch, and dull brown rather than chestnut brown flank and crown coloration. A three-note *chi-ca-go* call serves as a location call for both sexes.

Age and sex criteria. *Females* have dark brown rather than black crests and lack black throats.

Immatures have buff-tipped greater primary coverts for the first year, and their outer two primaries are relatively pointed and frayed.

Habitat and foods. California quail occur over a broad range of climates and vegetational types. Generally, however, they are found in areas of fairly mild winters and moderate precipitation during the breeding season, having an interspersion of brush, other vegetational types, and succulent plants or surface water. Nesting cover tends to be herbaceous rather than brushy, although the birds usually roost in tall shrubs or trees, and escape cover is preferably a heavy growth of surface vegetation into which the birds can easily run. Loafing cover that allows relief from the afternoon heat is desirable, and a reliable source of moisture is also needed. Foraging cover is variable since the birds eat a great array of seeds and leafy material, but leguminous plants are distinctly favored. Seeds provide the greatest part of the diet of California quails, with leafy material of secondary importance and grains, fruit, and insects of very limited significance.

Social behavior. The winter coveys of California quail are often relatively large, averaging about 30–40 individuals, but coveys of up to 500–600 birds have also been reported. Beginning in late winter, these coveys start to break up as unmated males separate from them to establish crowing perches. Pairs mated presumably from the past year remain in the covey until they

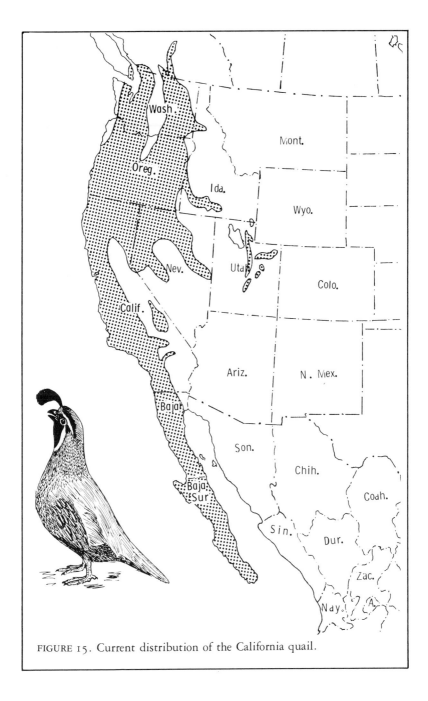

FIGURE 15. Current distribution of the California quail.

are ready to begin nesting. Unmated males utter their *cow* calls from various sites, frequently near mated pairs, and attempt to approach mated females whenever possible. Thus, if a mated male should be killed, the female is soon able to mate with one of the "crower" males. Likewise, when males lose their mates they begin crowing within a day. However, monogamy is the rule, and the male's role in achieving successful reproduction is an important one.

Reproductive biology. Nests are built in a large variety of protected situations; and although they are normally at ground level, they have also been observed in vines and even on rooftops. The eggs are laid at the rate of about 5 per week, and an initial clutch averages 13 or 14 eggs. Incubation is performed by the female and takes 22–23 days. On the loss of a female her mate will often take over the incubation. Broodless males also take great interest in young chicks and make excellent foster parents. It may be fairly common for the male member of a pair to take over the care of the first-hatched brood, enabling the female to begin a second. Parents and chicks are gradually joined by unsuccessful adults and later by unmated males after they stop their crowing behavior, eventually re-forming the coveys. Because of persistent re-nesting and possible double-brooding, productivity in these birds is fairly high, although the mortality rate of chicks between hatching and the fall hunting season may approach 50 percent.

Bobwhite

Colinus virginianus (Linnaeus) 1758

Other vernacular names. American colin, partridge, quail.

Range. Virtually all of the eastern United States north to southern Maine, New York, southern Ontario, central Wisconsin, and central Minnesota, west to southeastern Wyoming, eastern Colorado, eastern New Mexico, and eastern Mexico south to Chiapas and adjacent Guatemala, but excluding the lowlands of Yucatán. Also resident in Cuba and the Isle of Pines. Exists as isolated populations in Sonora (largely extirpated) and as introduced populations in the Columbia and Snake river basins of Washington, Oregon, and Idaho and in northwestern Wyoming (Bighorn and Shoshone river valleys). Recently reintroduced with questionable success into southern Arizona.

Identification. Length, 9.5–10.6 inches. The sexes are very different in appearance, and males vary greatly in coloration in different parts of the species' range. Males of most races have a white eye stripe that extends from the bill through the eye back to the

FIGURE 16. Current continental distribution of bobwhite.

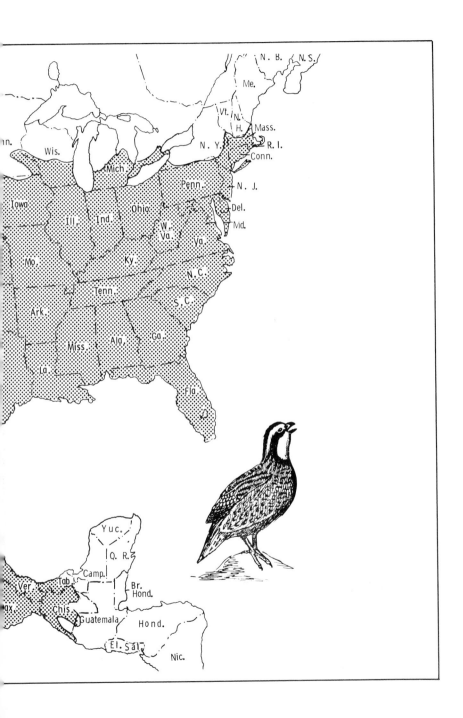

base of the neck, with brown to brownish black coloration above. The ear region is blackish to hazel brown in males, and this coloration extends backward below the white eye stripe and expands under the throat to form a blackish chest collar under the chin and throat, which are white in most races. However, in some populations the chin and throat are also black, and the lower chest may be either blackish or brownish. In the northern populations the male's breast and abdomen are irregularly barred with black and white, but in southern Mexico all under parts are generally darker and lack white markings. Females of all races have buff chins, upper throats, and eye stripes, and buff rather than white markings on the under parts. Females also lack black collars and in general are more heavily marked with brown and buff barring or mottling both above and below.

Field marks. Except in some parts of Mexico, the presence of a white throat and a white eye stripe that contrasts with an otherwise brownish to blackish head will serve to identify male bobwhites. Gray partridges might be confused with bobwhites, but the gray partridge has no white or pale buff on the head and also has a uniformly grayish chest. The whistled *bobwhite* location call of males in spring is distinctive, and similar whistled notes serve in reassembling scattered coveys.

Age and sex criteria. *Females* have buff chins and upper throats rather than white (or black in Arizona and some Mexican areas) like males. The whiter chins of males appear to some extent even in the juvenal plumage. The beak coloration (pale yellow at the base of the lower mandible in females; uniformly black in males) is useful in determining the sex of birds as young as 6–8 weeks old. The sex of birds at least 8 weeks old can be determined on the basis of the central portion of the middle wing coverts. Males have fine, black, sharply pointed and well-differentiated markings here, whereas females have wider, dull gray bands that do not contrast sharply with the rest of the feathers.

Immatures can often be identified by their outer two primaries, which are more pointed than the others, and the greater coverts of the first seven primaries, which have buff tips. A small percentage of birds may still be of questionable age by these two criteria, in which case first-year birds may be identified by using the seventh primary covert. It is usually brownish with buff tipping and is somewhat ragged, while in adults this feather is darker and sleeker, and has more whitish downy tipping at the base.

Habitat and foods. Over the very broad geographic range occupied by bobwhites in North America they are found in a great many habitat types, but a few common characteristics emerge. In all, a grassy cover for nesting, brushy cover for escape, and a source of cultivated crops or natural plant foods are needed. A great deal of interspersion of these habitat types is desirable as well. Unlike some other quails, bobwhites roost in low and fairly open cover where there is little obstruction of light and rapid escape can be made by flight. Nesting is usually in open herbaceous cover with a nearly bare surface and easy access from the ground. Bobwhites eat an extremely varied diet, but in general the plant component ranges from over 90 percent in winter to about 70 percent in summer, when insects are widely available. The most important plant items of diet are seeds of legumes and weedy herbs, and cultivated grains. A supply of water is more important to bobwhites than to most other American quails.

Social behavior. During the colder parts of the year the social unit is the covey, which usually numbers 10–15 birds. This is the most efficient number for forming their typical circular and heat-conserving roosting groups, with tails touching and heads facing outward. With the onset of spring, previously paired birds re-form pair bonds and unpaired males come into reproductive condition. The male's *bob-white* call is characteristic of such unpaired birds, and is rarely if ever uttered by mated birds, at least so long as their females are in view. Whistling males thus call attention to themselves and may succeed in attracting unmated females or those which have lost their mates. After pairing, the pair becomes quite inconspicuous and begins to seek out a nesting place.

Reproductive biology. Both sexes participate in nest building, which is initiated by the digging of a scrape. This is then filled with leafy material, and finally grasses or other herbaceous plants are arched over the top, effectively concealing the nest from above. The first egg is deposited in a day or two, and thereafter eggs are laid at a nearly daily rate until the clutch of about 14 eggs is complete. Incubation, which is performed by the female, requires 23 days. Males regularly undertake incubation if their mate is killed, and it is possible but still not adequately proven that males may at times take on the complete responsibility for rearing the brood after a week or so, leaving the female to attempt a second clutch. However, in most areas, the breeding season is not long enough to allow for double brooding, although as many as 2 or even 3 renesting attempts will be made if the early ones should fail. As the summer progresses, well-grown broods are joined by adults that were unsuccessful in breeding, and early fall covey sizes thus average about 12–17 birds.

Harlequin Quail

Cyrtonyx montezumae (Vigors) 1830

Other vernacular names. Black quail, crazy quail, fool quail, massena quail, Mearns quail, Montezuma quail, painted quail, squat quail.

Range. Southwestern United States south to Oaxaca, Mexico, primarily in pine-oak woodlands.

Identification. Adults, 8–9.5 inches long. The sexes are very different in appearance. Males have a beautiful facial pattern of black or bluish black and white and a soft, tan crest that projects backward and downward over the nape. The upper parts of males are grayish to olive brown, extensively spotted and marked with black, white, and buff. The sides and flanks are dark grayish, with numerous rounded spots of white, cinnamon, or reddish brown, depending on the population. The breast is unmarked brown, grading gradually to black on the abdomen and under tail coverts. Females are generally cinnamon-colored, with extensive blackish markings on the back. The female has a small, buff crest that is less conspicuous than the male's, and a mottled brown and buff face with a whitish chin and throat. The upper surfaces of the back and wings are extensively mottled, and the under parts are mostly buff with black flecks or streaks in the abdominal region.

Field marks. Males are unmistakable if their distinctively patterned face can be seen or if their extensively spotted flank pattern is visible. Females are more uniformly cinnamon-colored below than other species of quails. Unlike the scaled quail of the same region (which occurs in more open habitats), harlequin quail rarely run and instead tend to crouch and hide. They are seldom found far from pine-oak woodlands throughout their entire range. The distinctive call consists of a series of uniformly paced whistling notes, slowly descending in scale.

Age and sex criteria. *Females* lack the adult male's black and white ornamental patterning of the face and throat, having instead a white or buff chin and throat.

Immatures may be recognized by the greater primary coverts, which are edged with buff or barred near the base with buff (in adults these are spotted with whitish in males or barred with wide white markings). Also in immatures the outer two coverts are pointed rather than rounded. The condition of the outer primaries does not appear to be very useful in determining age.

Habitat and foods. The harlequin quail is perhaps more closely tied to a specific habitat than any other American quail, and it has the most restricted U.S. range of all. It is in general confined to pine-oak woodlands of the Southwest, although the critical habitat component is not the pines or oaks themselves, but the accompanying species of bulb-bearing forbs, especially wood sorrels and nut grasses. Provided that these succulent plants are present, the birds can probably get along without surface water, but their destruction by grazing or other means causes rapid elimination of the quail. Less important foods include acorns; seeds of legumes, grasses, and pines; and fruits of various shrubs

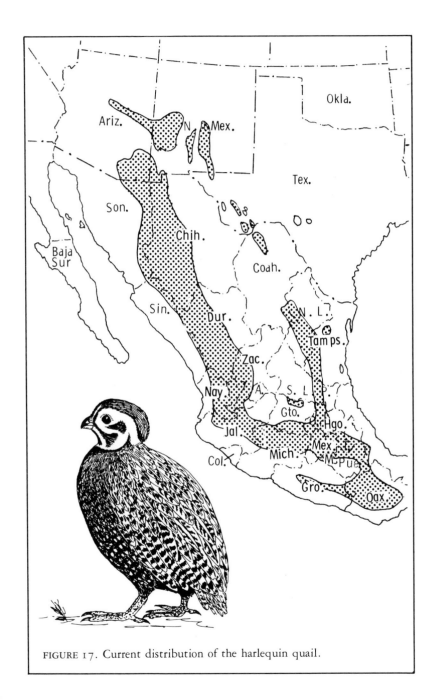

FIGURE 17. Current distribution of the harlequin quail.

and trees. Insects are eaten by adults during the summer and are relished by young chicks, but are not a major item of diet at any time after the earliest post-hatching stages.

Social behavior. During the nonbreeding season, harlequin quail form small coveys that probably represent family units. These average only 7 or 8 birds, and rarely are more than 25 ever seen. They typically feed in close groups, digging out the subsurface bulbs on which they depend so heavily, and at night they often form semicircular roosts around a rock or clump of grass. The nesting season in the U.S. range is relatively late, apparently coinciding with the summer rainy period, and calling by unmated males is most conspicuous in July and August. The mechanisms of pair formation are still unstudied in this species, but probably differ little if at all from those of the other American quails.

Reproduction biology. The participation of the male in nest building and incubation is still somewhat uncertain, but it is probable that he at least assists in nest building. The nest is a domed structure of grass stems that appear to be woven together, and is roofed over to form a chamber some 4 or 5 inches high. It also has a side entrance that is usually well hidden by a mat of grass steps, which hang down over the opening like a hinged door, falling back into place whenever the female enters or leaves. The average clutch is about 10–12 eggs, but the egg-laying rate in wild birds is still unreported. Males have been reported sitting on eggs in a number of cases, and have also been observed sitting beside incubating females. The incubation period is fairly long, lasting 25–26 days. During this time the male apparently assumes the major responsibility for defending the nest. When the young hatch they are fed insects, seeds, and bulbs by both parents, but they begin to forage extensively for themselves by the time they are 2 weeks old. There is evidently little mixing or merging of broods, and most fall coveys appear to be made up of family units.

Mountain Quail

Mountain Quail

Scaled Quail

Gambel Quail

Gambel Quail

California Quail

California Quail

Bobwhite

Bobwhite

Mearns Harlequin Quail

Mearns Harlequin Quail

Gray Partridge

Gray Partridge

Chukar Partridge

Chukar Partridge

Ring-necked Pheasant

Ring-necked Pheasant

Lesser Sandhill Crane

Lesser Sandhill Crane

Lesser Sandhill Crane

American Coot

American Coot

Common Gallinule

Purple Gallinule

Common Snipe

Band-tailed Pigeon

Mourning Dove

Mourning Dove

White-winged Dove

Gray Partridge

Perdix perdix (Linnaeus) 1758

Other vernacular names. Bohemian partridge, English partridge, European partridge, Hungarian partridge, Hun, Hunkie.

Range. Native to Europe and Asia but introduced into North America and now widely established in southern Canada and the northern United States (see distribution map). The North American population was probably derived from stock representing several different geographic races.

Identification. Adults, 12–13 inches long. The sexes are similar in appearance. Adults are tawny cinnamon on the head except for an uncrested buffy brown crown and ear patch. The breast and upper abdomen are a finely vermiculated gray which is interrupted by a chestnut brown horseshoe marking in males (smaller or absent in females), and the gray flanks are similarly interrupted by vertical chestnut barring. The upper parts are grayish to brownish, with darker mottling in the wing region and conspicuous white shaft streaks on the scapulars. The upper tail

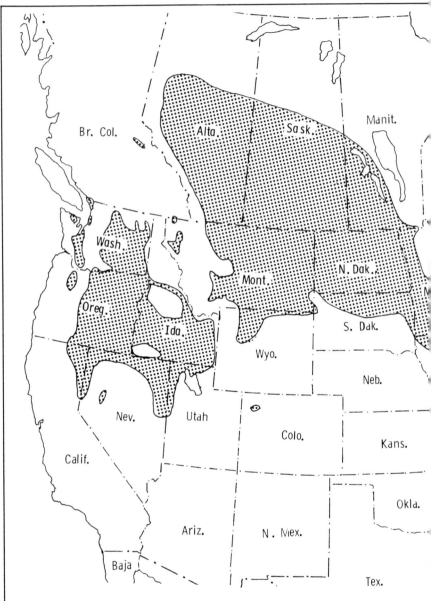

FIGURE 18. Current North American distribution of the
gray partridge.

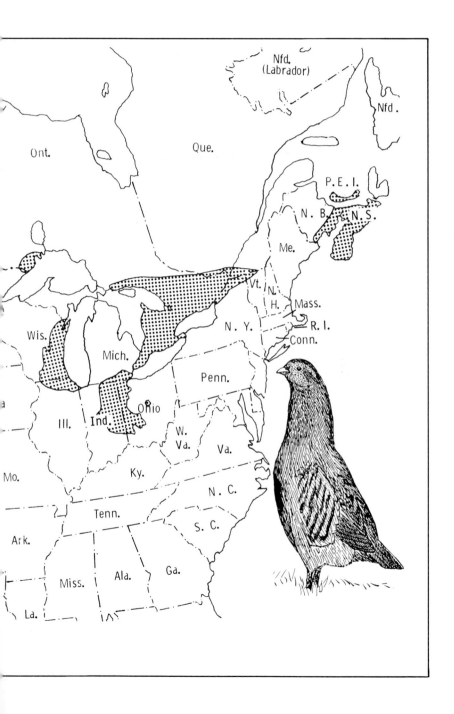

coverts and two central pairs of tail feathers are heavily vermiculated and barred, while the other tail feathers are rusty brown.

Field marks. In flight, the rusty tail feathers are spread and are usually conspicuous; otherwise, the impression is one of a grayish brown bird without bright markings. Chukar partridge also exhibit rusty outer tail feathers in flight but in addition have conspicuous white throats. The bobwhite occurs in some of the same regions as the gray partridge but is smaller and shows a grayish tail when flushed. In spring a raspy *tur-ip* call may be heard; it has also been described as a "rusty-gate" or *kee-uk* call.

Age and sex criteria. *Females* may sometimes but not always be identified by the scapulars and middle wing coverts, which may have a wide buff stripe along the shaft. The scapulars of males are yellowish brown with very fine wavy black lines running across each feather and a chestnut patch near the outside edge. Females have scapulars that are blackish at the base with about two light yellow crossbars, and only the outer parts of the feather are vermiculated.

Immatures have the usual pointed outer primaries and, at least for a time, have yellow rather than blue gray feet. In immatures the outer two primary coverts from the juvenal plumage are also retained; the ninth covert is typically pointed rather than rounded, and although it is like that of adults in being brown with white barring, it is only rarely tipped with white.

Habitat and foods. The gray partridge has adapted to a wide geographic range in North America and a variety of climatic conditions, but there is no clear pattern of association between its adopted range and vegetation or climate. Generally, however, the climates are those with fairly short growing seasons and limited precipitation during the incubation and brooding periods. The severity of the winter is apparently not a limiting factor, at least where grain or other seeds remain accessible. The species seems to favor combinations of croplands and native grasses:

woody and brushy cover types are evidently of little significance for its survival. Native grasslands or hayfields are used for nesting cover, and the same habitat type plus a source of water or succulent vegetation is needed by broods. The diet includes three primary types of foods: cultivated grains, seeds of various weedy herbs, and green leafy materials. Gray partridges eat all of the major small grain crops of the northern plains, especially corn, oats, barley, and wheat. Only during summer are insects eaten in any quantity, and even then in only limited quantities by adult birds.

Social behavior. As in native American quails, the social unit of the gray partridge is a moderately large covey of 10–20 birds, and rarely more than 30. Each fall covey is apparently produced by a pair with their surviving young, plus additional unsuccessful adult birds. The covey remains intact through winter, and pair formation evidently begins considerably before the spring break-up of the social unit. Previously paired birds often remate, but rarely if ever does pairing occur between broodmates or between parent and offspring. The "rusty gate" call, which sounds like *keee-uck!,* is the unmated male's principal advertisement call and is heard especially during early morning and evening hours. Pair formation seems to be a very gradual process, and potential pairs often change mates before settling into a permanent pair bond.

Reproductive biology. The female apparently builds the nest while the male stands guard. The nest is usually located in herbaceous cover, often in alfalfa fields, and is a shallow scrape lined with dead herbaceous vegetation. The eggs are laid at a rate of 1 egg in slightly more than 1 day, so that about 20 days are needed to complete an average initial clutch of 15–17 eggs. Incubation lasts for 24–25 days, and is apparently performed entirely by the female. Males remain close by, however, and at the time of hatching may sit beside the female on the nest. Both parents attend the generally large broods, but chick mortality is often high, especially when there is cold weather or rain during the hatching period. Although double brooding is not known to occur, persistent renesting usually results in a relatively large proportion of immature birds in the fall flocks. On the average, about 8 chicks per hatched brood survive to the start of the fall hunting season, giving the normal covey size of 10 or 12 birds for that time of year.

Chukar Partridge
Alectoris chukar (Gray) 1830

Other vernacular names. Chukor, Indian hill partridge, rock partridge.

Range. Native to Eurasia, from France through Greece and Bulgaria, and also southeastward through Asia Minor and southern Asia. These two populations should probably be regarded as separate species, and all of the introduced United States stock is apparently derived from the Asian species. The present range of the North American population is from southern interior British Columbia southward through eastern parts of Washington, Oregon, and California to the northern part of Baja California, and east in the Great Basin uplands through Nevada, Idaho, Utah, western Colorado, Wyoming, and Montana, with small populations of uncertain status in Arizona, New Mexico, western South Dakota, and southern Alberta.

Identification. Adults, 13–15.5 inches long. The sexes are identical in appearance, with white or buffy white cheeks and throat separated from the breast by a black collar or necklace that passes through the eyes. The crown and upper parts are grayish brown to olive, grading to gray on the chest. Otherwise, the under parts and flanks are buff, with conspicuous black and chestnut vertical barring on the flanks. The outer tail feathers are chestnut brown. The bill, feet, and legs are reddish, and males often have slight spurs on the legs.

Two other closely related species have been locally introduced into some western states and might occasionally be encountered. These include the Barbary partridge (*Alectoris barbara*) and the red-legged partridge (*A. rufa*). All have *chu-kar* calls and red legs, but the Barbary partridge has a reddish brown rather than black collar and a grayish throat and face terminating in a chestnut crown. The red-legged partridge more closely resembles the chukar partridge, but its black neck collar gradually blends into the breast by breaking up into a number of dark streaks, whereas in the chukar partridge the collar is clearly delineated from the grayish breast. Barbary partridges have been successfully introduced into California and red-legged partridges have been introduced without success into various states including Utah, Texas, and Colorado. They have possibly survived in eastern Washington.

Field marks. The striking black and white head pattern of this species can be seen for considerable distances in the arid country which it inhabits, as can the contrasting flank markings. In flight the reddish legs and chestnut outer tail feathers are usually visible. The *chu-kar* call often indicates the presence of this species.

Age and sex criteria. *Females* have no apparent plumage differences from males, and measurements must be used to distinguish the sexes. After the third primary is fully grown (by about 16½ weeks of age), the distance from the tip of the feather to the wrist joint (where the alula feathers insert) is over 136 mm. (5 $^5/_{16}$ inches) in males and under 136 mm. in females.

Immatures may be recognized by the fact that the length of the primary covert for the ninth primary is less than 29 mm. (2¼

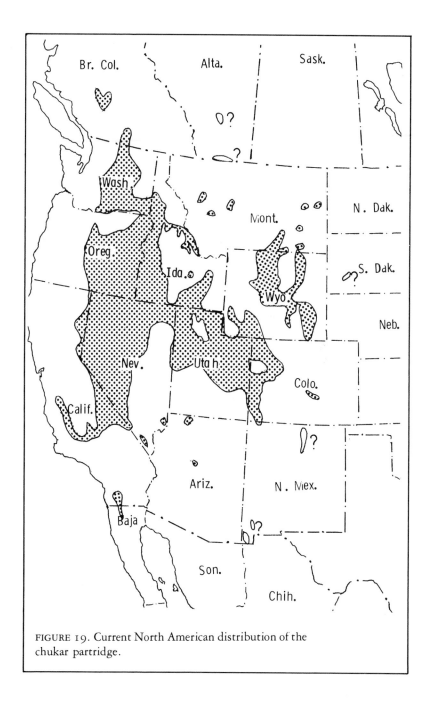

FIGURE 19. Current North American distribution of the chukar partridge.

inches) in immatures and is 29 mm. or more in adults. Since some chukars molt their ninth primary the first year, determining age from the outer primaries is often difficult, but in general the presence of faded vanes and pointed tips on the outermost or two outer primaries would indicate an immature bird. These feathers may also have a yellowish patch near the tip.

Habitat and foods. Throughout nearly all of the chukar partridge's adopted North American range, the typical vegetation is a sagebrush-grassland community, although in the extreme southern portions the species also occurs in a saltbrush-grassland community. The associated climate is one of short but hot summers and long, moderately cold winters. In addition, the species prefers rocky slopes, which allow for ready escape as well as shade and roosting cover. In such dry areas water during the summer months may be a limiting factor, and thus summer rains or a source of succulent plants during the breeding season are important. The birds feed primarily on the leaves and seeds of grasses, especially cheatgrass, and the seeds of a variety of weedy herbs. Although legume seeds are not especially important foods, the leaves of various legumes are sometimes eaten. Adults eat few insects or other animal material, but they are an important item of diet for growing chicks.

Social behavior. From the appearance of broods in late summer until the formation of pairs in the spring, the covey is the social unit of the chukar partridge. The size of coveys varies greatly according to population density and local conditions, but averages about 20 and infrequently exceeds 40 birds. By late winter pair formation is well underway, with the older birds pairing (or repairing with old mates) earlier than the younger ones. Although the species is basically monogamous, some observers have reported that a small proportion of males will pair with two females. The characteristic *chuck* or *chuckar* call not only is used as an advertisement display by unpaired males, but also serves as a signal by both sexes for reassembling scattered coveys, and is even uttered by paired males apparently to repel other males. Thus, the call functions in additional and rather different ways from the advertising calls of unmated males among American quails.

Reproductive biology. Nest building may be initiated by the male's "nest ceremony," in which he enters a clump of vegetation, crouches, raises and spreads his tail, and calls while making nest-scraping motions. Females sometimes perform the same ceremony before their mates, and thus both sexes may be involved with nest-site selection. Eggs are laid at the rate of an egg every day or two, and clutches vary in size from about 10 to more than 20 eggs. The incubation period is known to be 24 days, but there is still uncertainty whether the male regularly participates in incubation or brood care. Some observers suggest that males often desert their mates early in incubation and join other males, while others believe that males sometimes take over incubation on the first clutch, leaving the female free to begin a second one. The latter situation is much more likely, considering what is known about other species of quails and partridges. After hatching, there is a strong tendency for broods to intermix, perhaps because of frequent contacts between broods at watering places. This might also help account for the rather large ranges in observed fall covey sizes.

Ring-necked Pheasant
Phasianus colchicus Linnaeus 1758

Other vernacular names. Chinese pheasant, chink, ringneck.

Range. Native to Asia but introduced extensively into North America and now widely established. Currently found in North America from the Queen Charlotte Islands and southern British Columbia southward in the Pacific coastal states to the Imperial Valley and extreme northern Baja California, locally in the intermontane region along river valleys or in irrigated areas of Idaho, Nevada, Utah, Arizona, and New Mexico, and in the grasslands east of the Rocky Mountains from central Alberta eastward across Saskatchewan, southern Manitoba, southern Ontario, New York, New Hampshire, Maine, New Brunswick, and Prince Edward Island to Nova Scotia, and southward to New Jersey, Pennsylvania, Ohio, Indiana, Illinois, Missouri, Kansas, western Oklahoma, northern Texas, and New Mexico.

Identification. Adults, 21–25 inches long (females) or 30–36 inches long (males). The adult male ring-necked pheasant is almost impossible to confuse with any other species; its long, pointed and barred tail distinguishes it from all other North American species except the sage grouse, and the latter has feathered rather than spurred legs and lacks the pheasant's iridescent coloration. Female pheasants have a relatively long and strongly barred tail too, but lack spurs and are a dull mottled brown color throughout. They are about the same size as the female sage grouse, but the latter have a dark abdomen patch and feathered legs, while female ring-necked pheasants are buff on the under parts and have bare legs. Sharp-tailed grouse might perhaps be confused with female pheasants, but the formers' tails are much shorter and they also have feathered legs.

Field marks. Pheasants are likely to be found in open grassland and cropland areas where some brushy cover also exists, and unless pressed are more likely to run than to fly. On takeoff, the males often utter a croaking call, and the long, pointed tail of both sexes is distinctive. Except during the nesting season, males are frequently seen in company with a harem of several females. In the spring the male's territorial call, a loud, double-noted *squawk* followed by a softer wing-whirring sound, can often be heard.

Age and sex criteria. *Females* can normally be recognized readily by the absence of spurs, iridescent coloration, or extensive bare red skin around the eye. Some old females or those with damaged ovaries may assume a rather malelike plumage, but they lack spurs.

Immatures closely resemble adults by their first fall of life, and unlike most North American gallinaceous birds do not retain the two outer juvenal primaries through the winter. The presence of growing or recently replaced outer primaries thus indicates a young bird. Young males may usually be distinguished from adults on the basis of their spurs, which are lighter in color, usually blunter and not decurved, and softer and less glossy than in adults. The bursa of Fabricius is a reliable indicator of age in

FIGURE 20. Current North American distribution of the
ring-necked pheasant.

first-fall birds: its maximum length is 6 mm. (¼ inch) in adult females and 8 mm. (⁵/₁₆ inch) in adult males, while younger birds have a longer bursa.

Habitat and foods. The pheasant's winter habitat, at least in the colder parts of its range, must offer adequate cover extending above snow line and a source of food. These conditions are met in marshes, plum thickets, shelter belts, and heavy brush in ravines and along fence rows or railroad right of ways. Grain such as corn or milo can provide a supply of food even if it must be scratched out from under a foot or two of snow. The ideal spring habitat consists of a diversity of cover types that provide food, escape cover, and nesting sites. Fields of alfalfa, sweet clover, or small grains, and fence rows, are favored nesting sites. Roadside ditches, particularly those that have an abundance of early-maturing rather than warm-season grasses, are also valuable for nesting. Brooding habitats must have an abundance of insects, edible green vegetation, and adequate escape and roosting cover.

Social behavior. During the winter, mixed or single-sex groups of pheasants congregate in areas of food and cover, but by early spring the males begin to disperse and establish "crowing areas." These areas are not typical territories and have indefinite boundaries, but by his crowing and wing-whirring displays the male may attract a harem of several females. A "waltzing" display, in which the male lowers the wing closer to the female and moves past her, is one of the typical pheasant displays. After fertilization the female leaves the male's company to establish a nest, which may or may not be within the area originally defended by the male.

Reproductive biology. Nests are simply a scooped-out depression in the soil to which a lining of feathers and plant materials is gradually added during egg laying. Eggs are laid at the approximate rate of one per day until the clutch of a dozen or more is complete. Compound clutches, or "dump nests" resulting from the efforts of more than one female, are not uncommon. Incubation begins with the laying of the last egg, and requires approximately 23 days. The entire clutch hatches almost simultaneously and the newly hatched young may leave the nest only a few hours after hatching. The female attends her brood throughout their juvenile period, usually for from 6 to 8 weeks. By their eighth week of life, young males are beginning to show their distinctive breast coloration, but molt in the young birds continues through the fifth month of life. Both adult females and males also molt during this period. By fall the adults and young begin to gather in fields of ripening grain, from which they gradually move into heavier cover as winter begins.

Cranes, Rails, Coots, and Gallinules

(Order Gruiformes)

Lesser Sandhill Crane

Grus canadensis canadensis (Linnaeus) 1758

Other vernacular names. Little brown crane.

Range. Breeds from the west coast of Alaska and adjacent Siberia eastward across arctic Canada, including Banks and Victoria Island, to Keewatin, Southampton Island, and the western coast of Hudson Bay, and southward to the Aleutian Islands, Cook Inlet, and across northern Canada possibly as far south as the Hudson Bay watersheds of western and northern Ontario. Intergradation with the poorly defined Canadian race *rowani* evidently occurs in the northern parts of Alberta, Saskatchewan, and Manitoba, making the southern breeding limit of *canadensis* impossible to define accurately. Winters in the southwestern United States and Mexico, principally in eastern New Mexico and northwestern Texas.

Identification. Adults 35–40 inches long. One authority reports that lesser sandhill cranes seldom have a bill length (exposed culmen) greater than 3⅞ inches, or 100 mm., while the Canadian race *rowani* seldom has a bill length less than that. The bill length of greater sandhill cranes is never less than 121 mm. (4¾ inches) in adult males and is seldom less than 120 mm. in females. Weight appears to be a less useful criterion for distinguishing these subspecies, apparently because of dietary or seasonal variations. In all subspecies the adults are grayish, variably stained with brown during the breeding season.

Field marks. Sandhill cranes of all subspecies are so tall and long-legged that they can only be confused with other cranes and with the larger herons or egrets. The larger whooping crane differs in having a white plumage with black wing tips; even the more brownish immature whooping cranes have appreciably whiter plumages than do sandhill cranes. The great blue heron is similar in proportions and general coloration to sandhill cranes but has a yellowish bill with a black and whitish crested head, while the sandhill crane has a black bill and a reddish crown color. The lower back of the crane has a tuft of down-curved feathers that hide the tail, and in flight cranes hold the neck outstretched, in contrast to the folded neck postures of herons and egrets. The bugling *karoo* call of sandhill cranes carries for great distances.

Age and sex criteria. *Females* cannot be distinguished from males by external criteria, although they average slightly smaller in size. Internal examination of the sex organs is thus required for certain determination of sex.

Immatures carry tawny-colored feathers from the juvenal plumage through the first fall and into the first winter of their life, but by the spring after hatching most of these feathers have been replaced by the mouse-gray body feathers of the older age classes. The young birds have a high pitched *peer* call until they are about 10–12 months of age, when they acquire the bugling adult call. Since sandhill cranes evidently do not molt their tail feathers during their first fall or winter, it is possible that young birds might be recognized until their second summer of life by the presence of notched tail feathers, although this has not been specifically

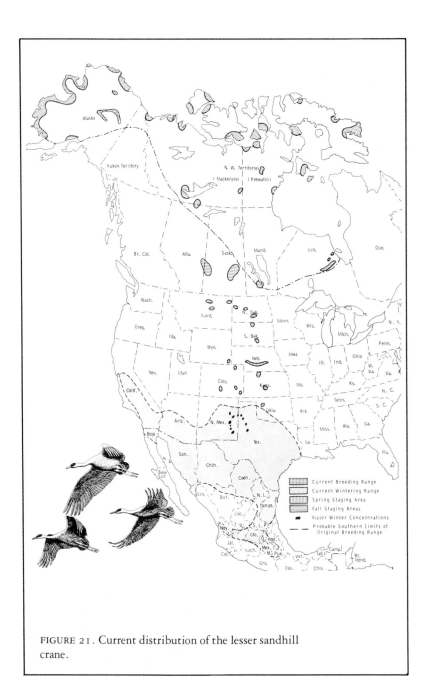

FIGURE 21. Current distribution of the lesser sandhill crane.

noted. It is reported that in addition to the obvious plumage and vocal differences between adults and young, juvenile birds during their first fall have hazel-colored rather than reddish eyes, a wine-cinnamon rather than drab gray bill color, and a uniform gray-brown head coloration rather than the contrasting red and pale gray pattern of older age classes.

Habitat and foods. The Alaskan breeding habitat of this strongly migratory subspecies consists of low-marshy tundra and, to a lesser extent, brushy muskeg, grass- and sedge-dominated marshes, and even occasionally sandy knolls and dunes. During the fall, migrant flocks concentrate in wheat and corn fields in Canada and the northern states, where they feed on the ripening grain. In the wintering areas of Texas and New Mexico the birds concentrate on the grassy and arid Staked Plains, where shallow lakes or rivers provide sandbars and islands for roosting sites. From these roosts the birds fly out daily to forage in newly sprouted wheat or wheat stubble, in alfalfa, and on such grain crops as sorghum, corn, and barley. Grasshoppers and probably other insects are also eaten when they are available. During the spring migration in the Platte Valley of Nebraska the birds forage primarily in wet meadows and corn stubble, eating waste corn, new volunteer corn shoots, and various succulent plants native to the meadows. They roost on the islands and sandbars of the wide and shallow Platte.

Social behavior. While sandhill cranes are in winter flocks, and especially just before the nesting season, they often engage in spectacular "dances." Pairs, and sometimes larger groups of birds, perform a series of vertical bounding movements, with wings partially spread and head bowed. Sometimes cornstalks or other bits of vegetation are flung up during these bouncing dances. The functions of dancing, which is common to all cranes, are still obscure, but it is not restricted to pair formation or the mating season. Cranes also perform the display when disturbed or when nesting birds are threatened. Threat displays toward other cranes involve a stiff-legged strutting toward the opponent, with

neck arched, bill pointing downward, and inner secondary feathers raised. Display preening often precedes an actual attack, which includes wing beating, clawing, and bill stabbing.

Reproductive biology. After sandhill cranes mature, which may require four years, they remain paired for life. Established pairs defend the same territory year after year, and may even use the same nest site. The nests are rather simply constructed and vary from an inconspicuous accumulation of grass and twigs in a dry location to a large haystacklike structure built near the edge of a marsh. Both sexes defend the nest and territory, and both take turns at incubation, with the female usually incubating at night. Incubation of the clutch of 2 eggs (rarely 1 or 3) begins with the first egg and probably requires 30 days for hatching, judging from knowledge of the greater sandhill crane. The eggs usually hatch a day apart, and in many cases one chick (or "colt") will be tended by each parent. Unless the young are raised separately, antagonism between them often results in the intimidation and starving of the smaller bird. Thus, most pairs of cranes raise only one offspring per year. The fledging period is not certain, but probably is less than the 65–75 days reported for the more southerly nesting greater sandhill crane. Shortly after fledging of the young the families gather into large flocks, and the long southward migration begins.

American Coot
Fulica americana Gmelin 1789

Other vernacular names. Mudhen, rice hen.

Range. Breeds from central British Columbia eastward through southern Mackenzie, Saskatchewan, Manitoba, western and southern Ontario, southern Quebec, southern New Brunswick and Nova Scotia, and southward locally through the western states and Mexico to Nicaragua, and along the Caribbean and Gulf of Mexico northeastward to Louisiana, Mississippi, Tennessee, Indiana, Ohio, Pennsylvania, and Maryland, with local or sporadic breeding in various southwestern states (Florida, North Carolina, Alabama). Also breeds in the West Indies (Cuba, Jamaica, Isle of Pines, and Grand Cayman). Winters from the Pacific coastal, southern, and southeastern states southward through Mexico and Central America to Panama.

Identification. Adults, 13–16 inches long. Coots may be distinguished from all of the true waterfowl (Anatidae) by their scalloped rather than webbed toes. They differ from such near relatives as gallinules in having a white bill, black head feathers, and an otherwise mostly dark grayish body.

Field marks. On the water, coots somewhat resemble ducks, but have a conspicuous short, white bill and white under tail markings. They often feed in large groups near the shore or on it, or may dive in fairly shallow water for their food. They swim with the same head-pumping movements typical of gallinules. In flight their long legs and feet trail behind the tail, and the white tail markings as well as a white trailing edge of the secondaries contrast with the otherwise dark gray body coloration.

Sex and age criteria. *Females* can sometimes be identified by the length of the metatarsus (unfeathered portion of the leg) and middle toe, excluding claw (127 mm. [5 inches] or less, as opposed to 128 mm. or more in males), and by the length of the culmen and frontal shield, or distance from the tip of the bill to the top of the frontal shield (36.7–44.6 mm., or under 1¾ inches, as opposed to 43.2–49.7 mm., or over 1¾ inches, in males). If the middle toe (including the claw) and the metatarsus measure 140 mm. (5½ inches) or more, the bird is most probably a male; a shorter measurement is typical of females.

Immatures can be recognized by their blue or grayish green legs. Presumed second-year birds have yellow-green legs, third-year birds clear yellow, and in very old birds the legs may be reddish orange.

Habitat and foods. To a greater degree than any other of the North American Rallidae, the coot is closely associated with fairly deep ponds and relatively open water throughout the year. Its ecology is thus more like that of the diving ducks than of the typical rails, and like some diving ducks it eats largely submerged vegetation obtained by diving. It also feeds on shorelines, both on vegetable matter and, at least during summer, to a moderate extent on insects and aquatic animal life. During migration and in wintering areas the distribution of coots is broad and limited mainly by the availability of submerged aquatic

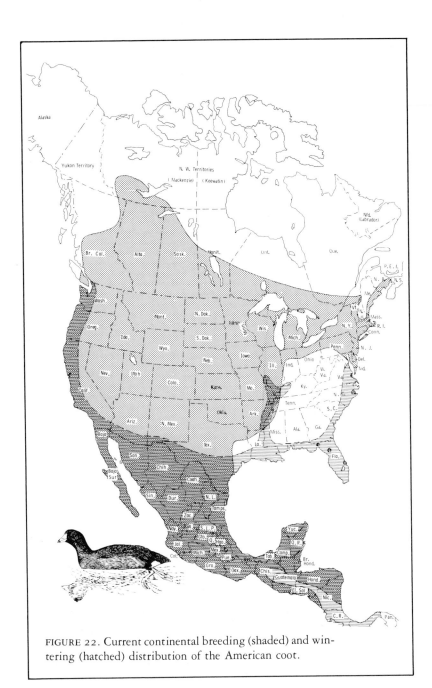

FIGURE 22. Current continental breeding (shaded) and wintering (hatched) distribution of the American coot.

plants at suitable depths for foraging. In coastal areas brackish estuarine bays are favored wintering habitats. Requirements for the establishment of breeding territories include emergent nesting vegetation such as cattails and bulrushes, as well as open water for foraging and territorial patrolling.

Social behavior. The life and social activities of coots revolve largely around territorial defense during their entire adult life. The birds are evidently wholly monogamous, with a long courtship period and a lifelong pair bond, although new mates may be acquired rapidly after the loss of a partner. Success in obtaining a mate is associated with success in acquiring and holding a territory. Small winter territories, or core areas, are held by males on the wintering grounds, where courtship occurs. If the birds winter on their breeding grounds, this core area may simply be expanded to form a nesting territory; otherwise, a nesting territory is established after arrival on the breeding area. This larger area (an acre or more in size) must include emergent vegetation suitable for providing nesting material and anchoring the floating nest. Within this area the male constantly patrols his territorial borders, performing a series of aggressive displays and attacking intruding birds. The species has an array of calls for individual recognition, courtship, alarm, perturbation, warning, and intimidation. A cocked-tail display following fights with other coots is typical, while the usual aggressive response toward other species is a spread-wing "swanning" posture. Courtship displays include billing, nibbling, and bowing. An arching display by the female initiates copulation.

Reproductive biology. During their reproductive cycle, coots build a variety of structures, including nestlike display platforms on which copulation occurs, nests for incubating their eggs, and brood nests. Males normally bring nesting materials to the female, who constructs the nest. Eggs are laid at nearly daily intervals, and clutches vary in size from 5 to almost a dozen, depending on season and location. Incubation begins before the clutch is complete and both sexes participate, with the male doing the majority of it. The incubation period is 23 days, and the chicks hatch over several days. Typically the female leads the first of the chicks to hatch away from the nest and begins to feed them, while the male remains behind to care for the later hatchlings. After 8 or more have hatched, the male usually leaves the nest, abandoning with his part of the brood any unhatched eggs. The young are initially fed by the parents, but soon begin to forage for themselves and after a month or so are nearly independent. They begin to fly at about 75 days of age, but may occasionally resort to begging food from their parents nearly up until that time. As the chicks mature, the parents become increasingly intolerant of them, and at times expel them from their territory even before they are fledged. In some areas a second clutch is begun after the first brood has dispersed. The age of sexual maturity is still rather uncertain, but at least some females are known to have bred at a year.

Gallinules

Gallinula and *Porphyrula*

Other vernacular names. Blue peter, moor hen, mud hen, pond chicken, water hen.

Ranges. *Porphyrula martinica* (Linnaeus): Purple gallinule. Breeds from the Gulf states and South Carolina southward to Peru, Uruguay, northern Argentina, and Brazil, and in the West Indies. Winters from Texas, Louisiana, and Florida southward.

Gallinula chloropus (Linnaeus): Common gallinule. In North America breeds locally in California and central Arizona, and from Nebraska, Iowa, Wisconsin, Michigan, southern Ontario, southwestern Quebec, and Massachusetts southward to Chile and Argentina. Winters from California and Arizona southward. Also breeds in Great Britain (where it is called the moor hen) and throughout Eurasia south to Ceylon and the East Indies.

Identification. Adults, 12–15 inches long. Gallinules can be distinguished from coots, their only near relatives, by the absence of scalloped edges on their toes, and from all other water birds of comparable size by their brightly colored and pointed bills, which merge with the frontal shields. Purple gallinules are more iridescent and have a bluish white frontal shield, while common gallinules are more blackish and have a bright red shield. In addition, purple gallinules have entirely white under tail coverts but lack white flank markings, while common gallinules have white flank markings and black central under tail coverts.

Field marks. Like coots, gallinules swim with a distinctive front-to-back head-pumping motion that distinguishes them from true waterfowl of similar size. They also wade, forage along shorelines, and may even climb low bushes. In flight, the purple gallinule exhibits a purplish upper body color and dangles its yellowish legs, while the common gallinule is grayish above and has greenish legs. Unlike coots, they show no white line at the rear edge of the secondary feathers during flight. Grating, croaking, or cackling calls are typical of both species of gallinules.

Age and sex criteria. *Females* are not readily distinguished from males by their plumage characteristics, although they average slightly smaller in their major measurements. Internal examination is required for certain determination of sex.

Immatures of both species lack bright bill coloration and are a duller and more brownish color than adults. Since coots are known to have a well-developed bursa of Fabricius during their first fall, it is probable that gallinules do also.

Habitat and foods. A species of the tropical and subtropical swamps and marshes, the purple gallinule is especially common where floating-leaf plants such as water lilies are abundant. In the Southeast, the birds are common in swamps having growths of pickerelweed, in which they often nest. They eat both plant and animal foods, including the seeds of rice and other grasses and aquatic insects, snails, and worms. During the autumn, rice is a favored food, and sometimes the birds cause damage to rice fields. In contrast, the common gallinule has a much more temperate distribution and breeds on or near still waters, from lakes to small ponds, and from small, slowly flowing streams to large rivers. Small water areas with abundant emergent vegetation and peripheral cover are preferred for breeding. This species eats mainly the seeds and fruits of weeds and grasses, although it also occasionally feeds on insects, earthworms, slugs, and snails, and, rarely, aquatic vertebrates like tadpoles and small fish.

Social behavior. Few detailed observations have been made of wild gallinules, which are both relatively shy and usually found in rather inaccessible habitats. Like other species of gallinules, they are known to be monogamous, highly territorial during the breeding season, and relatively aggressive during nonbreeding periods, when some flocking occurs. Vocalizations are an important part of their social communications, and both sexes have an astonishing variety of notes. Behavior between pairs of common gallinules early in the breeding season centers largely on nestlike display platforms, one of which may later be converted into a nest. Males repel intrusions into their territory by other gallinules of either sex with threats or attack; and after pair bonds are formed, both mates participate in expelling intruders.

Reproductive biology. At least in common gallinules both sexes take part in nest building, but it is chiefly done by the male. In the purple gallinule a well-developed and carefully constructed runway is built from the water to the nest and is invariably used by adults. The common gallinule sometimes makes a less elaborate runway. The clutch size in both species is usually about 6–8 eggs, laid on consecutive days (at least in the common gallinule). Incubation may begin with the first or second egg, or sometimes not until the clutch is complete. Incubation is shared by both sexes, in several-hour shifts, and the changeover may be marked by the presentation of a reed or leaf to the incubating bird, which adds it to the nest before leaving. Incubation lasts 19–22 days in the common gallinule and about 22 days in the purple gallinule. Depending on when incubation began, the young may hatch all at the same time or at intervals over a period of up to a week. The newly hatched chicks are left in the nest initially, and the parents remove the eggshells and eat the droppings of the chicks. They are fed bill-to-bill by the parents, mainly insects and other small animals. After a few days the chicks leave the nest and closely follow the parents, who continue to feed them for several weeks. One or more brood nests may be built in the brooding area and used by the family. By about 5 weeks of age the young can largely feed on their own, and they can fly in 6 or 7 weeks (common gallinule). Multiple brooding is the general rule, and 3 or perhaps more broods may be raised where the breeding season is long enough. The young of previous broods remain in the vicinity of the adults, and have even been reported to help care for later broods.

Purple Gallinule

Rails

Rallus and *Porzana*

Other vernacular names. Marsh hen, ortolan (sora), railbird, stage driver (king rail).

Ranges. *Rallus elegans* Audubon: King rail. Breeds in fresh-water marshes from North Dakota, Minnesota, Wisconsin, Michigan, southern Ontario, and New York south to Texas and Florida, as well as in the West Indies and central Mexico. Winters in the southern parts of its breeding range.

   Rallus longirostris Boddaert: Clapper rail. Breeds mostly in salt-water marshes from California south to Ecuador and from Connecticut south to Brazil, as well as in the West Indies and parts of central Mexico. Resident over most of its range.

   Rallus limicola Vieillot: Virginia rail. Breeds in fresh-water marshes from the southern half of Canada southward to southern South America. Winters mainly from the southern United States southward.

   Porzana carolina (Linnaeus): Sora. Breeds from northern and central Canada southward to northern Baja California, Nevada,

New Mexico, Missouri, Ohio, West Virginia, and Pennsylvania. Winters from the southern United States southward to South America and the West Indies.

Identification. Adults, 8–20 inches long. Rails differ conspicuously from other gray or brownish wading and swimming birds by their vertically barred flank plumage patterns. The four species that are usually shot as gamebirds and are included here differ considerably from one another in their overall body size and bill length. The sora has a bill less than an inch (24 mm.) long, the Virginia rail's bill is under 1⅝ inches (42 mm.) in length, and the king and clapper rails have bill lengths in excess of 2 inches (55 mm.). The latter two species can usually be discriminated by the paler and more grayish plumage of the clapper rail and its distinct salt-marsh habitat preferences.

Field marks. Rails of all species are relatively secretive, and usually are seen only when flushed at close range. After a short, weak flight they quickly drop back into cover and are hard to locate again. They produce a variety of sounds: the king rail a tubalike grunting, the clapper rail a wooden clattering, the Virginia rail a metallic clicking or squealing, and the sora a pulsating whinnylike sound. At close range the sora appears to be mostly grayish, and has a stubby yellow bill. The Virginia rail is the same general size (less than 12 inches long) but has a longer and rather reddish bill and a more brownish body color. The king and clapper rails are much larger, with yellow to brownish bills. The upper parts of the clapper rail are tinged predominantly with ashy gray and of the king rail with tones of reddish brown to buff.

Age and sex criteria. *Females* of most rails are almost identical in plumage to males, so weight and measurement data are usually needed for external sexing. Adult female sora rails, however, are said to have a gray auricular (ear) patch that extends unbroken from the eye to the gray nape region, while in males it is usually discontinuous with the gray of the nape and may or may not ex-

Sora Rail

tend to the eye. In males the stripes above the eyes are also usually broken by the black crown, while in females they are typically continuous. During the breeding season the male's bill is usually chrome yellow from tip to base, with a narrow white band around the base, while the female's bill normally shades to olive green at the tip and lacks the white band. Methods for externally sexing Virginia rails have evidently not been developed. Adults of both clapper rails and king rails can probably be fairly reliably sexed by weight. According to one researcher, males of these two species weighed at least 300 and 330 grams (10.5 and 11.6 ounces), respectively, while no females attained these weights.

Immatures of all rail species tend to be more uniformly colored and generally are a darker sepia or olive brownish tone than adults. Juvenile sora lack the black facial patterns of adults, and in addition retain their juvenile wing feathers during the first year. White barring or spotting on the upper wing coverts is probably typical of most young rails. The presence of a bursa distinguishes immature king rails, at least to early winter, and perhaps is useful in determining the age of the other rail species too. Mature king rails are distinguished from immature ones by having orange-yellow color at the base of the bill, a pinkish-brown heel mark, and orange-red rather than yellow tongue and mouth coloration. It is probable that the reddish bill and leg coloration of adult Virginia rails is likewise diagnostic of maturity, as are the yellowish bill and orange-yellow heel mark of the clapper rail. It is reported that by the age of 10 weeks, clapper rails differ from adults only in the duller color of their fleshy parts.

Habitats and foods. All of the rails are marshland birds, the clapper rail characteristic of coastal marshes and the other three of inland, fresh-water marshes. Although rails can swim, they typically forage from a standing position along the edges of heavy cover or where floating plant debris permits easy walking. Their ability to compress themselves until they are "thin as a rail" allows them to move unseen through heavy emergent vegetation. Soras eat mainly seeds, while the Virginia rail feeds heavily on insects. The king rail prefers crustaceans and aquatic insects for its primary diet but eats substantial amounts of rice in some wintering areas. Clapper rails have a more exclusively invertebrate diet and even in winter eat relatively little food of plant origin.

The nesting habitat of sora rails consists of small-grass- and sedge-dominated swales, ponds, and lake borders where the water is 6–8 inches deep and food is abundant. Virginia rails nest in much the same areas and habitats as soras, but their nesting locations are more varied and include sedges, cattails, and bulrushes, amid any of which their nests may be built. The clapper rail's nesting habitat is primarily cord grass marsh, especially where ditches or creeks cause an interspersion of tall and short grasses. King rails apparently have the most diverse habitats of these species, ranging from salt-water to fresh-water marshes and swamps, and even upland fields. They are, however, closely associated with muskrats, along whose water pathways they nest, obtain water, and forage for crayfish.

Social behavior. Evidently all of these rails are monogamous, with yearly establishment or reestablishment of pairs in conjunction with territorial defense. Upon returning to the breeding areas, males establish and proclaim their territories, sometimes the same as in previous years, by uttering their distinctive mating calls (the king rail's *kik-kik-kik,* the clapper's *keck-keck*, the Virginia's *ticket-ticket,* the sora's *whee-hee-hee* or whinny). In Virginia and clapper rails territorial defense is directed primarily against other members of the same species, but in king rails and soras it includes eviction of all other rails. Once mating has occurred, both sexes probably participate in territorial defense. Rails have a variety of additional calls associated with pair-bond maintenance and nesting. Pair-forming displays in the king rail consist largely of the mating call and flashing of the white under tail coverts, while courtship feeding of the female by the male is evidently an important pair-maintaining display. In the sora and Virginia rails apparently only the latter performs courtship feeding, but pairing and copulatory behavior in both include bouts of preening, standing-lateral displays, precopulatory chases, and postcopulatory movements. Mutual preening and symbolic nest building by the male have been observed in certain rails as well, and may occur in all of these species.

Reproductive biology. At least in the king rail, the nest site is probably chosen by the male, who may initiate its construction. It is usually provided with both a canopy and a ramp for easy access from water. The male king rail takes the major role in building the nest, which may not be finished before the first egg is laid. All four species also construct several additional brood nests, usually without canopies, near the egg nest. Their nests are usually in fairly uniform stands of vegetation, typically near the edge of heavy cover and a source of water. In all species the eggs are laid on an approximate daily basis. The sora begins incubation with the first few eggs that are laid, but other species do not initiate it until the clutch is nearly complete. Clutches usually number from 8 to 12 eggs, but as many as 18 have been seen. Thus, hatching may be staggered over as long as 17 days in soras, while in the others it is usually completed in 1 or 2 days. Both sexes incubate and defend the nests, feigning injury when threatened and sometimes attacking the intruder. The incubation period is 18–19 days for the sora and Virginia rails, and 21–23 days for the larger species. Both sexes brood the newly hatched young, and in the sora the female sometimes takes the first-hatched birds and leaves the male to attend to the remainder. The chicks are fed bill-to-bill by the adults, but as they begin to feed for themselves their begging is ignored by their parents, who may eventually begin pecking them and driving them from their territory. This behavior allows the start of a second brood in areas where the breeding season is long enough. Second broods have been proven for the clapper rail in South Carolina, and there is evidence that the other three species undertake them too. The young of clapper and king rails are able to fly at 9 or 10 weeks of age, while in sora and Virginia rails adult proportions, and presumably also flight abilities, are attained by 6 weeks of age. During the brooding period the adults undergo their postnuptial molt and become flightless for a time, but young birds retain their juvenal flight feathers through the first winter.

Woodcocks and Snipes

(Order Charadriiformes)

American Woodcock
Philohela minor (Gmelin) 1789

Other vernacular names. Timberdoodle, wood snipe.

Range. Breeds from southeastern Manitoba eastward through southern Ontario, southern Quebec, New Brunswick, Prince Edward Island, Cape Breton Island, and southern Newfoundland, southward through Minnesota, eastern Iowa, and locally to Missouri, Arkansas, and Louisiana, and eastward to central Florida. Most breeding probably occurs from the Great Lakes region northward, and eastward through the Allegheny Mountains. Winters from eastern Texas and eastern Oklahoma eastward through Arkansas, Louisiana, and Mississippi, with smaller numbers wintering through the southeastern coastal states northward to Virginia and, occasionally, New Jersey.

Identification. Adults, 10½–11½ inches long. Woodcock are distinguishable from all other North American birds on the basis of their long (over 2 inches) bill, rotund body, and the large, high-placed eyes. Unlike the common snipe or other brownish shore birds, the woodcock has a series of blackish bars extending across rather than along the crown. In both sexes, but especially in males, the three outer primary feathers are unusually narrow.

Field marks. Woodcock are rarely seen except in flight, since they remain "frozen" until the last possible moment before flushing. As they take off, their wings produce a whistling sound, and the combination of a long bill and short tail is distinctive. In spring, the males' nasal *peent* calls can be heard at dusk, as they get ready to perform their aerial display. In flight display the birds hover on whistling wings, trilling a series of notes that sound like *chickaree, chickaree.*

Sex and age criteria. *Females* are readily distinguished by their bill length, the width of the outer three primaries, or a combination of these traits. All birds with a bill length of 72 mm. (2¾ inches) are females, and all adult birds with a bill of less than 64 mm. (2½ inches) are males. The combined width of the three outer primaries (measured 2 cm., or ¾ inch, from their tips) in males totals 12.4 mm. (½ inch) or less, while in females it totals 12.6 mm. or more. By the start of the fall hunting season, females usually weigh over 175 grams, or 6 ounces, and males less than that.

Immature birds may be readily identified by examining the under surface of the secondary feathers. They have a light-colored band at the outer edge and a darker bar just below it. In adults the dark bar is lacking. Immature birds also have a bursa of Fabricius.

Habitat and foods. The habitats used by woodcocks through the year must include locations for singing, nesting, brood rearing, and migration and wintering. Throughout their range and the year's cycle woodcocks are generally confined to young forests with scattered openings on poorly drained land. The male's spring singing ground must have a clearing in which the bird can fly, with the size of the clearing directly related to the height of the surrounding trees. Plant life in this clearing is typically of the

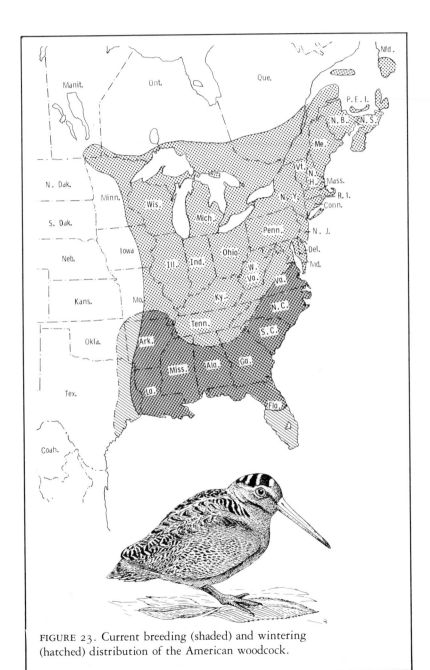

FIGURE 23. Current breeding (shaded) and wintering (hatched) distribution of the American woodcock.

early woody or low shrubby stage, with some herbaceous cover for alighting. The nest is usually built fairly near the edge of woody cover, and may be in mixed hardwood and conifer growth, in pure hardwood, especially alder, cover, or in old fields or brushy cover. It ordinarily is within a few hundred feet of a singing ground. Brood cover is very similar to nesting cover, and must be close to poorly drained areas where the chicks can begin probing for worms. During migration and through the winter both adults and young forage primarily on earthworms, and are restricted to habitats high in this food source. Shrubby habitats having a herbaceous ground cover support large numbers of earthworms, and thus also of woodcocks, especially where soil moisture is adequate but not extreme and soil texture favors moisture retention.

Social behavior. Although some of the literature is contradictory on this point, it seems that woodcocks are polygamous if not promiscuous. Males apparently do not assist with brood rearing, and will usually attempt to mate with decoys placed in their singing fields. After their spring arrival on the breeding grounds, males establish territorial singing grounds, from which they exclude other males and at which they perform dawn and dusk song flights. After a long series of *tuko* and *peent* calls, the male launches himself into the air, rising in a series of diminishing circles until he is hovering high above the ground. Then, as his wing "twittering" ceases and he begins a zigzag glide toward the ground, he utters a series of liquid trilling notes. Females are attracted to these singing grounds, and copulation occurs there. Females also utter the *peent* and *tuko* notes, but do not perform display flights. A limited amount of display occurs after the breeding season, and some song flights are initiated on the wintering grounds in December and January.

Reproductive biology. The relatively rudimentary nest of the woodcock is built entirely by the female, often in sparse cover at the base of a shrub or small tree, or near the edge of a shrub thicket. The clutch normally contains 4 eggs, although some late clutches may have only 3. The eggs are laid at a daily rate, and the incubation period of 20–21 days begins after the last egg is laid. During incubation the female is a notoriously "tight" sitter, relying on her effective camouflage to avoid detection. She feigns injury when forced from a hatching nest; and shortly after the chicks are hatched, they "freeze" when threatened. A brooding female performs a labored flight when flushed with her brood, and observations of the related European woodcock suggest that she simulates the carrying of a chick between her legs by sharply depressing and fanning her tail feathers. The young make their first flights when they are about 2 weeks old, and can fly well by 3 weeks. Broods probably break up some 6–8 weeks after hatching, and apparently the birds gradually migrate southward at low altitude in loosely scattered flocks. Evidently most woodcocks migrate to traditional wintering areas each year and return to their original breeding grounds for breeding.

Common Snipe
Capella gallinago (Linnaeus) 1758
(Often classified as *Gallinago gallinago*)

Other vernacular names. English snipe, jacksnipe, Wilson's snipe.

Range. Breeds extensively in Europe and Asia from Great Britain to Siberia. In North America breeds from Alaska eastward across Yukon, the Northwest Territories, Manitoba, Ontario, Quebec, and Labrador, and south to southern California, Utah, northern Colorado, Iowa, northern Illinois, Ohio, Pennsylvania, and New Jersey, with sporadic or local breeding farther south (Arizona, southern Colorado, Nebraska, Oklahoma). Winters from southern British Columbia and Maryland southward through Mexico, Central America, and northern South America (Colombia and Brazil).

Identification. Adults 10–11½ inches long. In the hand, common snipes can be distinguished from other brownish shore birds of comparable size and proportions (dowitchers, lesser yellowlegs, etc.) by their strongly striped crown pattern and generally

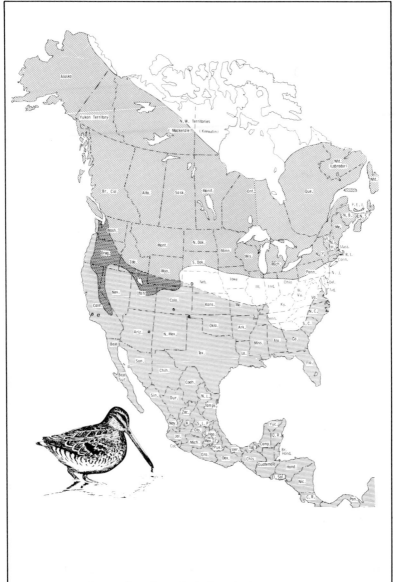

FIGURE 24. Current breeding (shaded) and wintering (hatched) North American distribution of the common snipe.

rust-colored tail, which has slightly narrowed and mostly white outer feathers, especially the outer pair.

Field marks. Like woodcock, common snipes are rarely seen until they fly, but usually flush from wet meadows, marshes, or boggy areas rather than from wooded swamps or alder thickets. As they flush, they utter a raspy *scaipe* note, and fly off in a low, erratic manner before dropping down into the marshy cover again. On the breeding grounds, the aerial display is marked by a "winnowing" *huhuhuhuhuhu* sound made as the bird performs a nearly vertical "power dive" with its wings half-folded and its tail widely spread, causing the outer pair of feathers to vibrate. At the end of the dive the bird pulls out, regains altitude, and repeats the performance.

Age and sex criteria. *Females* are reportedly identical to males in plumage, so internal sexing may be necessary. So far, no reliable external sexing or aging methods are available.

Immatures cannot be reliably distinguished from adults after they are fully grown, at least by plumage criteria, but the length of the bursa of Fabricius (at least 3 mm., or ⅛ inch, in young birds) may be a useful aging guide.

Habitat and foods. The common snipe's wintering habitat includes a variety of marsh types—delta and prairie marshes, rice fields, and cattail thickets—all characterized by wet organic soils. Marshes are also used during migration, but breeding usually occurs in bogs or low tundra vegetation. The bird's diet is about 80 percent animal material, primarily insects as well as smaller amounts of crustaceans, earthworms, and mollusks. They also eat seeds like those of sedges, smartweeds, bulrushes, and other marsh plants. Most of the food is obtained by probing in shallow water, but some surface feeding is also apparently done.

Social behavior. Even during winter, snipe are apparently not especially gregarious, but may concentrate in areas of open water during freezing weather. At times flocks are seen in migration at a fairly high altitude, but they are not nearly so conspicuous as many shore birds, and probably migrate mainly at night. Shortly after they reach the breeding grounds, the males establish territories and await the arrival of the females 10–14 days later. The most common male territorial display is "bleating," or "winnowing," an aerial performance in which the male (and later also the female) rises some 300 feet into the air, then performs a series of swooping dives at the rate of about 8 per minute. In these dives the wings are partially closed and the tail feathers, especially the outer pair, are spread wide. The resulting interruption of air flow from the wings by the vibrating outer feathers produces an eerie tremolo sound. Territorial birds also have several vocalizations, including a repeated *cut-a-cut* "yakking" call, which is usually uttered on both ascent and descent during the aerial display flight, and on the ground as well. Several other ground calls and displays are typical, including a tail-fanning display for distracting intruders' attention from the nest. Common snipe are monogamous, and the male remains with his offspring through the brood-rearing period.

Reproductive biology. The nest is usually built in wet, marshy ground, especially where low grass clumps or brush rises above the surface of the bog, and is simply a depression in such a clump or tussock, lined with grasses or dead leaves. The clutch size is nearly always 4 eggs. Incubation, performed entirely by the female, normally takes 18–19 days. Typically the first 2 chicks that hatch are adopted by the male, while the female cares for the latter 2, and the parents raise their chicks separately. The chicks are fed by the parent, who probes in the mud and allows them to eat as the bill is withdrawn. Chicks make their first short flights at the age of about 15–18 days. By the time they are about 6 weeks old they begin to form groups ("wisps") with others of their own age, and sometimes such flocks may number in the hundreds. These young birds apparently migrate south together in advance of the adults, and it is probable that adult females migrate south somewhat before adult males.

Pigeons and Doves

(Order Columbiformes)

Band-tailed Pigeon

Columba fasciata Say 1823

Other vernacular names. Blue pigeon, blue rock, white-collared pigeon.

Range. Breeds from the middle of western British Columbia southward along the Pacific coastal states to the Mexican border and slightly beyond into Baja California. Also breeds in mountain forests from central Utah and northern Colorado southward through Arizona, New Mexico, and southwestern Texas into Mexico to the Isthmus of Tehuantepec, and below the isthmus in the mountains of Chiapas, Guatemala, Honduras, Nicaragua, Costa Rica, and Panama. Winters from central California, central Arizona, and New Mexico southward.

Identification. Adults, 13–15 inches. Band-tailed pigeons are the only North American pigeons with a white crescent marking on the back of the neck and a rounded tail with a broad pale gray band at the tip. The feet are yellow and the bill is yellow tipped with black, while the plumage is mostly bluish gray above and ruddy brown ventrally.

Field marks. Band-tailed pigeons might possibly be mistaken for domestic pigeons (rock doves), which are of about the same size and body proportions, but band-tails have yellow feet and a pale-tipped tail, while rock doves have red feet and a black-tipped tail. Band-tailed pigeons are also typically found in mountain forests rather than urban or suburban areas. The advertising call of the band-tailed pigeon is rather weak cooing, usually an *oo-whoo* that is uttered five times, with the last two calls shorter and dropping sharply in pitch.

Age and sex criteria. *Females* have a more brownish or grayish breast color, while males' breasts are bluish. This difference provides a 95 percent accurate sex-determination character by the time the birds are only 8 weeks old. The neck crescent in males is wide and white, whereas in females it is thin, mottled, and pale. Further, the upper wing coverts of adult males have a light blue metallic hue, but in females they are a much darker gray, with only a faint trace of iridescence.

Immatures can be readily identified by the buff tips on their greater wing coverts; about 20 percent of the juveniles retain such coverts after the last juvenile primary is lost at about 8 months of age. Also, about 60 percent retain one or more rounded juvenile secondaries (the fourth, fifth, or sixth) after the last juvenile primary is replaced.

Habitat and foods. In the United States, band-tailed pigeons are associated with the western mountain oak woodlands or woodlands of mixed piñon pine and oaks of the arid southwestern mountains. Heavily wooded canyon slopes provide typical nesting habitats. In Colorado, migrant birds prefer stands of Gambel oak and ponderosa pine, especially if situated near fields of small grains, while higher coniferous forests are used for nesting. An abundant food supply, especially of the favored acorns, is a major determinant of distribution patterns. A variety of other natural nuts (including pine nuts) and berries are also prime foods. During winter and into spring the band-tailed pigeon feeds largely

FIGURE 25. Current distribution of the band-tailed pigeon.

on acorns, but during summer it eats a variety of cultivated crops (oats, cherries, peas, and wheat). As fall approaches, the birds return to a diet of nuts, fruits, and berries. Since they are quite gregarious, even during the nesting time, they may consume a substantial amount of food and may cause local damage to crops. At various seasons flocks gather at sources of mineral water or even at salt deposits, for dietary reasons that are still uncertain.

Social behavior. Up until the dispersal of pairs for nesting, the flock is the social unit for these birds. Even paired birds may return at times to feed with the flock, and there is an immediate reformation of flocks after the fledging of the young. As birds that are already paired (presumably from the past year) disperse for nesting, unmated males begin to establish "cooing perches" in trees, from which they call and launch out into circular display flights that carry them over potential nesting areas. Females that enter the male's display area are evidently recognized by their failure to respond aggressively, and the male then begins his courtship behavior. This includes cooing, a head-swinging display, preening behind the wing, and attempts to mount the female in a strutting posture. After pairing, "nuptial flights" involving a circular glide over the nesting area while calling are continued, and evidently function as territorial displays. Groves of pines or oaks commonly comprise the center of the nesting territory, within which roosting and all activities other than feeding occur.

Reproductive biology. Although the nest is built by the female, the male probably selects the nesting tree. These are usually located near a clearing, often on a slope or precipice, and the nest is normally built on a horizontal branch several feet from the trunk and less than 25 feet from the ground. When a nest site is selected, the male brings twigs to the female, who constructs the bulky nest. After completion of the nest, which may require from 2 to 6 days, the single egg is laid. Incubation usually takes 19 days, with the male incubating from morning until late afternoon and the female through the night. The nestling is brooded by both parents for the first 19 or 20 days, and fledging occurs about 4 weeks after hatching. At least in California, up to 3 broods per season may be reared. As the last nestlings fledge, families begin to band together into small flocks, which probably are later reconstituted into larger groups.

Mourning Dove

Zenaida macroura (Linnaeus) 1758
(Regarded until 1973 as *Zenaidura macroura* by the A.O.U.)

Other vernacular names. Turtle dove, wild dove.

Range. Breeds from southern British Columbia eastward through the southern part of central Canada to southern Quebec, southern New Brunswick, and adjacent Nova Scotia southward through the entire continental United States, plus nearly all of Mexico except the Yucatán peninsula, and extending southward through Central America to Panama. Also breeds in the West Indies. Winters from the southern half of the United States southward at least as far as Panama.

Identification. Adults, 11–13 inches long. Mourning doves are the most widespread and familiar of the dove family in North America, and the sole surviving native species with a sharply pointed tail. The only white present is on the tips of the lateral

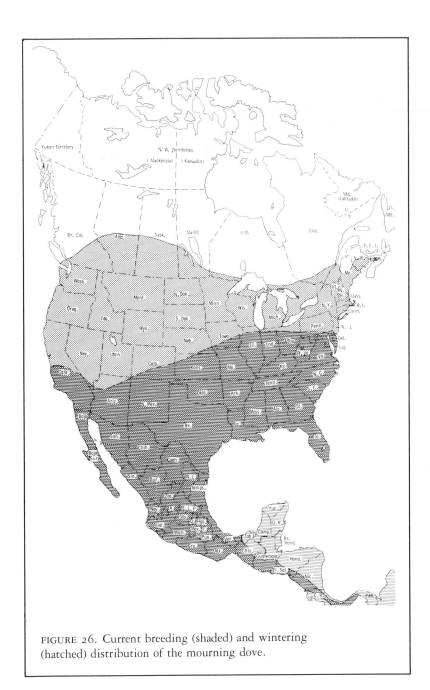

FIGURE 26. Current breeding (shaded) and wintering (hatched) distribution of the mourning dove.

tail feathers; most of the plumage is grayish brown. The black ear mark and faintly iridescent neck plumage are like those of the white-winged dove, but this species has extensive white markings on the upper wing surface and lacks exposed black spots on the scapulars and wing coverts.

Field marks. When the birds are in flight, the pointed wings and sharply pointed tail are the best field marks. The flight is direct, the wing beats are swift, and the wings make a slight whistling sound in flight. The call is a mournful *ooah, coo, cooo, coo,* with the last three notes the loudest and carrying the farthest.

Age and sex criteria. *Females* are very similar in plumage to males, but whereas the males have a faintly pink breast and a bluish gray crown, these areas in females are more brownish. The neck iridescence is less prominent in females, as is the ear patch, and the black spots on the scapulars and wing coverts are larger and more numerous.

Immatures (up to about 5 months) can be recognized by the whitish tips of the upper primary coverts and by their retention of one or more outer juvenal primaries. After that age, young doves can be identified with certainty by the presence of a bursa of Fabricius.

Habitat and foods. This species is the most widely distributed and ecologically tolerant of the North American doves and pigeons, breeding in a broad range of habitats, from desert areas close to water to a variety of forested environments. It not only tolerates but benefits from human conversion of natural habitats into cultivated areas or pasturelands. It is especially abundant in fields, orchards, or generally weedy areas having an abundance of grains or seeds, which constitute its primary foods. Cultivated grains like corn and wheat are eaten wherever they are available, and seeds of doveweeds, bristle grass, ragweed, pigweed, and other annual weeds are all locally important items of diet. Other than a source of food, breeding habitat needs are very flexible. Trees or shrubs that provide horizontal nest platform supports are used for nesting when available, but in some areas of southern California the birds regularly nest on the ground. A supply of water is desirable but not necessary, since the birds can go several days without drinking. This enormous adaptability for nesting habitats and persistent renesting or multiple nesting are no doubt important aspects of the species' success.

Social behavior. Like other species in the dove family, mourning doves are strongly monogamous, with pair-forming displays beginning shortly after the migrant birds' arrival on the breeding grounds. Males quickly establish territories, from which they engage in display flights that involve towering upward with noisy wing beats, then gliding downward on outspread wings, often in a graceful arc. The defended territorial area includes potential nest sites but not food or water supplies. The distinctive advertising call of territorial males is familiar to almost everyone, consisting of a very faint double cooing note followed by two or three much louder ones. With the appearance of an unpaired female, the male utters his "display coo," similar to the advertising call but perhaps given with a slightly greater intensity. In his display before the female, he inflates his neck while cooing, exhibiting his iridescent neck feathers, and rhythmically spreads and closes his tail. This display evidently occurs when the male is demonstrating at a potential nest site. An upward flicking of the tail signals an intention to fly.

Reproductive biology. Shortly after pairs are formed, a nest site is selected, normally with the male taking the initiative in locating potential sites. Calling and caressing between the pair at the selected site may be important in maintaining pair bonds. Pairs may construct their own nest, but also frequently use the old nests of other birds, such as robins or grackles, or an old nest of their own. In such cases of nest re-use, new material is added to the nest before the eggs are laid. The 2 eggs that make up the total clutch are usually laid about 24 hours apart. An incubation period of 14 days is most common, although some nests require 15 or even 16 days to hatch. As in other species, males normally incubate during the day and the female at night. Hatching of the 2 eggs more frequently occurs on successive days than on the same day. However, at times the younger bird may be the more aggressive in obtaining food from its parents and may even grow more rapidly than its sibling. Young usually leave the nest on the twelfth day after hatching, and they are fully fledged at 13–15 days. In cases of renesting after a successful rearing of a brood, courtship may begin when the young are 9 days old and the first egg of the new clutch may be laid when the young are still only 10 or 11 days of age. Thus, pairs typically build a new nest for their second clutch, and may alternate between the two nests for successive nesting attempts. Since as few as 25 days may elapse between the start of one successful nesting and the laying of the first egg of the next brood, and it is apparent that multiple brooding occurs in most areas, the rearing of four broods is not uncommon. As the young of each brood becomes independent, they gather in small flocks and later concentrate in favored foraging areas. Social dominance hierarchies, similar to pecking orders in chickens, may develop in wintering flocks, and the resulting dominance rank influences both mating patterns and subsequent reproductive success.

White-winged Dove
Zenaida asiatica (Linnaeus) 1758

Other vernacular names. Mexican dove, Sonora dove, white-wing, white-winged pigeon.

Range. Breeds from southeastern California, extreme southern Nevada, southern Arizona, southwestern New Mexico, and southern Texas southward through Mexico, including the Baja California peninsula, and Central America to Costa Rica and western Panama. Also breeds in the West Indies. Winters mostly in its breeding range, except for the northern parts.

Identification. Adults, 11–12 inches long. This is the only native or introduced North American dove that is extensively marked with white on the upper wing surface. Apart from this trait, and the species' more bluntly tipped tail, it closely resembles the similarly grayish brown mourning dove. In the white-winged dove the tail is less than two-thirds as long as the folded wing, while that of the mourning dove is at least three-fourths the length of the wing.

Field marks. White-winged doves can be identified by their white wing markings both when perched and when in flight. The rounded, rather than pointed, tail distinguishes them from mourning doves, but otherwise they have the same general body outline in flight. The call of the white-winged dove is distinctive: a harsh, repeated cooing that sounds like "who cooks for you," with the last note most strongly accented.

Age and sex criteria. *Females* exhibit more grayish brown and less purplish on the crown and back of the neck than do males, and are less iridescent on the sides of the neck. They also have a smaller ear spot and browner under parts than males.

Immatures may be recognized by their pink to salmon-colored legs, as opposed to the distinctly reddish feet of adults, and they also have primary coverts with pale tips. The plumage of immatures is generally much grayer than that of adults, and the neck lacks iridescence.

Habitat and foods. The United States population of white-winged doves includes two subspecies, which inhabit somewhat different terrains and nesting habitats. However, both types of habitat fall into the general category of semiarid woodlands, with densely foliaged trees of low to medium height and fairly open ground cover. In Texas these include woodlands of mesquite and acacia, as well as mature citrus groves in recent years. Nesting habitats for the more arid-adapted western subspecies include chaparral, oak woodland, desert grassland, and desert scrub. In all habitats the birds feed primarily on seeds, mast, and fruit, although cultivated crops like sorghum, wheat, barley, corn, rice, and beans are also major foods in most regions. Doveweeds and other species of the spurge family are favored natural foods, while seeds of various composites, cactus fruits, prickly poppies (*Argemone*), legumes, and grasses may be locally important items of diet. A reliable source of drinking water during the hot summer months is also an important habitat component.

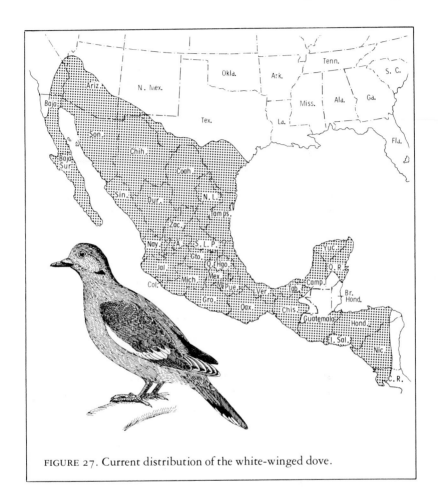

FIGURE 27. Current distribution of the white-winged dove.

Social behavior. In this migratory and monogamous species, courtship apparently begins during spring migration or no later than arrival at the breeding grounds. Males establish territories within which nesting, roosting, courting, and mating occur. A central component of the territory is thus a roosting and nesting location, which is usually a thorny tree but may be a large cactus. Territorial advertisement consists of display flights and calls— two, including one that lasts about 3 seconds and another (or extended version of the same) that lasts up to about 20 seconds. When a female joins the male on his territory, intensive courtship display begins, followed by nest building. Territories are defended against intruding mourning doves as well as other white-winged doves.

Reproductive biology. Although the male evidently selects the general nesting site, his mate chooses the exact location for the nest, usually a horizontal limb, crotch, or other flat surface. The male selects nesting materials and carries them to the female, who incorporates them into the nest. Most nest building is done in the morning hours, and it may take from 2 to 10 days to complete. The 2 eggs that make up the clutch are laid at an interval of approximately 36 hours. When, on rare occasions, more than 2 eggs are found in nests, they may be the work of additional females. Both sexes share the incubation equally, with the male sitting during most of the daylight hours and the female from late afternoon until morning. Incubation lasts 14 days, and the eggs usually hatch about 24 hours apart. Feeding is done by both parents, who regurgitate a thick fluid called "pigeon milk" in the usual manner of doves. The young leave the nest after about 2 weeks, at which time they are able to fly fairly well. Multiple broods are typical, with 2 broods usually being raised successfully, and as many as 5 nesting attempts may be made. The young of the first brood are allowed to remain in their parents' territory but are excluded from those of neighboring pairs. It is still uncertain how long family bonds persist during the fall.

Keys to Identification

The three following keys can be used to identify unfamiliar species of American game birds other than waterfowl that may be examined in the hand. Unless one is certain that the bird represents a particular taxonomic group (order or family) of birds, he should begin with the first key. The procedure, as in the use of all such keys, is to choose whichever of the two initial descriptive couplets (A or AA) best fits the unknown bird. Having chosen one of these, proceed to the choice of couplets (B and BB) occurring immediately below the chosen couplet, without further regard for descriptions listed below the rejected alternative. After making a varying number of such choices in the first key, the reader will have identified the bird as to its major taxonomic group. If it belongs to either the galliform or the gruiform group, the following two keys may be used to identify the specimen as to its species. These two keys operate in the same fashion as does the first, by the reader again starting with the choice of couplets A and AA and proceeding until the bird has been identified as to species. Measurements, where they are given, refer to adult birds, but in general the keys have been devised in a manner that will allow for identification regardless of the specimen's sex or, within limits, its age.

Key 1: Major Groups of North American Game Birds

A. Front toes fully webbed, swimming and diving birds with bills having toothlike projections on margins of both mandibles . . . ducks, geese, and swans of family Anatidae, not included in present work.

AA. Front toes not fully webbed, nor with bill as described above.

    B. Bill short and chickenlike; land birds with well-developed front toes and a weaker hind toe that is usually elevated . . . gallinaceous birds of order Galliformes; see Key 2.

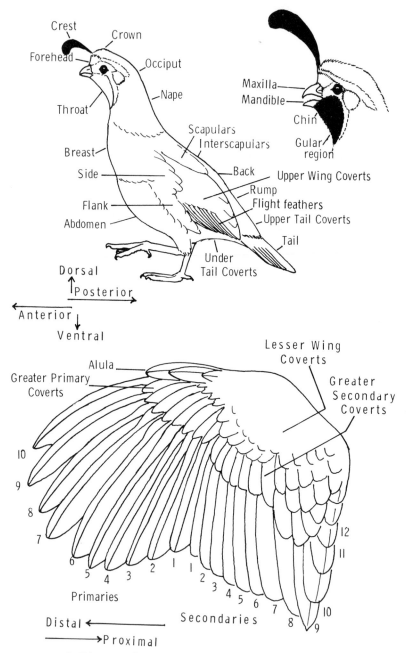

FIGURE 28. Diagram of general feather regions of game birds (above) and of upper wing surface (below).

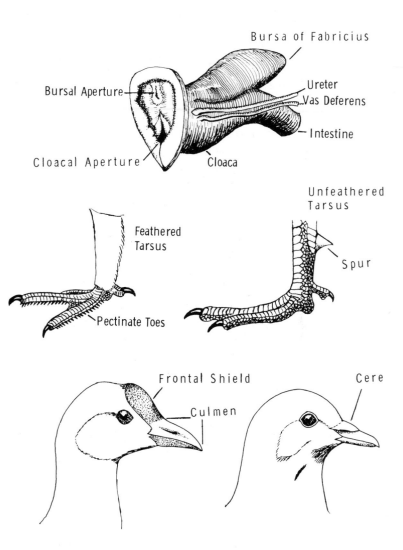

FIGURE 29. Diagram of the bursa of Fabricius (above), foot characteristics (middle), and bill characteristics (below) of game birds.

BB. Bill long and not distinctly chickenlike; perching, swimming, or wading birds with hind toe variably developed.

   C. Hind toe less than half the length of the lateral toes, wading and swimming birds with variably elongated bills that lack a fleshy enlargement (cere) above the nostrils.

      D. Bill variable in length but usually sharply pointed, with the ends not pitted and the tip of the lower mandible not covered by the upper one . . . cranes, coots, gallinules, and rails of order Gruiformes; see Key 3.

      DD. Bill long and tubular, with a flexible tip that is flattened and pitted, and with the upper mandible covering the tip of the lower one . . . snipes and woodcocks of family Scolopacidae. See text for distinction of the two species included here.

   CC. Hind toe more than half the length of the lateral toes, perching birds with weak bills that have a soft and fleshy enlargement (cere) above the nostrils . . . doves and pigeons of family Columbidae. See text for distinction of the three species included here.

Key 2: Identification of Galliform Game Birds of America

A. Hind toe not elevated and more than half the length of the lateral toes . . . chachalaca of family Cracidae.

AA. Hind toe elevated and less than half the length of the lateral toes . . . family Phasianidae.

   B. Head and upper neck naked, larger birds weighing over 9 pounds (3,000 grams; . . . turkey of subfamily Meleagrinae.

   BB. Head and upper neck feathered, smaller birds weighing considerably less than 9 pounds (3,000 grams).

      C. Lower leg (tarsus) largely or entirely feathered, the

toes also feathered or with comblike (pectinate)
margins present . . . grouse and ptarmigans of
subfamily Tetraoninae.

D. Tail feathers (rectrices) all sharply pointed,
larger birds usually weighing over 3 pounds
(1,100 grams) . . . sage grouse.

DD. Tail feathers not sharply pointed, usually
with squarish tips, smaller birds usually
weighing under 3 pounds (1,100 grams).

E. Lower half of tarsus unfeathered, sides of
neck having broad, ornamental "ruff"
feathers . . . ruffed grouse.

EE. Tarsus feathered to base of toes or
beyond, neck feathers not as described
above.

F. Outermost tail feathers less than $^4/_5$
length of central ones, outer webs of
primaries regularly patterned with
white or buff spots.

G. Central pair of tail feathers considerably longer and different
in color from others . . .
sharp-tailed grouse.

GG. Central pair of tail feathers not
markedly different from others,
neck with tapered, erectile pinnae . . . pinnated grouse.

FF. Outermost tail feathers more than
$^4/_5$ length of central ones, outer
webs of primaries irregularly mottled or uniformly colored.

G. Upper tail coverts not extending to tip of tail.

H. Tail feathers 16 (rarely
18), under parts heavily
barred . . . spruce
grouse.

HH. Tail feathers 18–20 (rarely 16), under parts mostly grayish . . . blue grouse.

GG. Upper tail coverts extending to tip of tail, which usually has 16 feathers.

H. Lateral tail feathers white . . . white-tailed ptarmigan.

HH. Lateral tail feathers dark brown or black.

I. Bill black and heavier (usually over $1/4''$ high at base), folded wing over $7.5''$ (195 mm.) long . . . willow ptarmigan.

II. Bill grayish at base and slighter (usually under $1/4''$ high at base), folded wing under $7.5''$ long . . . rock ptarmigan.

CC. Tarsus unfeathered, toes never feathered or with comblike margins.

D. Cutting edge of lower mandible usually with one or more slight identations, tarsus never with a sharp spur, with from 10 to 14 tail feathers . . . New World quails of subfamily Odontophorinae.

E. Tail less than half the length of the folded wing, a bushy crest of soft, broad feathers present at nape, tips of extended feet reach beyond tail . . . harlequin quail.

EE. Tail longer than half the length of the folded wing, virtually crestless or bearing a distinct crest near front of head, tips of extended feet not reaching the end of tail.

    F. Back feathers (scapulars) and inner wing feathers (tertials) spotted, tail less than $2^3/4$ inches (70 mm.) long . . . bobwhite.

    FF. Back feathers and inner wing feathers not spotted, tail over $2^3/4$ inches long.

        G. Crest of two narrow, black plumes, folded wing over $4^3/4$ inches (120 mm.) long . . . mountain quail.

        GG. Crest not as above, wing under $4^3/4$ inches long.

            H. Crest bushy and buff-colored, body feathers marked with dark scallops . . . scaled quail.

            HH. Crest of brown or black feathers that curve forward and are enlarged toward their tips.

                I. Abdomen feathers edged with darker color in a scalloped pattern, flanks marked with olive brown . . . California quail.

                II. Abdomen feathers extensively blackish, or buff with mottling or

streaking, flanks
marked with chestnut
. . . Gambel quail.
DD. Cutting edge of lower mandible not in-
dented, with or without spur on tarsus, 14
or more tail feathers . . . pheasant and
partridges of subfamily Phasianinae
E. Larger birds (usually over 2 pounds, or
900 grams), tarsus usually spurred, tail
long and pointed . . . ring-necked
pheasant of tribe Phasianini.
EE. Smaller birds, tarsus unspurred, tail
short and rounded . . . partridges of
tribe Perdicini.
F. Throat white, 14 tail feathers . . .
chukar partridge.
FF. Throat brown, 16–18 tail feathers
. . . gray partridge.

Key 3: Identification of Gruiform Game Birds of America

A. Bill about 4 inches (100 mm.) in length, crown area
mostly naked and reddish in adults, plumage predominantly
grayish . . . sandhill crane of family Gruidae.
AA. Bill less than 4 inches long, crown area always feathered
. . . rails, coots, and gallinules of family Rallidae.
B. Forehead with a bare frontal plate, flank feathers never
vertically barred.
C. Toes with scalloped edges formed by lateral lobes,
bill white . . . American coot.
CC. Toes with narrow lateral edges, bill red and yellow.
D. Frontal plate on forehead bluish white, body
plumage mostly purplish . . . purple gallinule.
DD. Frontal plate on forehead red, body plumage
mostly slate gray . . . common gallinule.

BB. Forehead without a bare frontal plate, flank feathers
with vertically barred feather pattern.
>   c. Bill under 1 ³/₄ inches (42 mm.), smaller birds
>   weighing less than 5 ounces (140 grams).
>   >   D. Bill short, stubby, and yellow, face and throat
>   >   partly black in adults . . . sora.
>   >   DD. Bill longer (longer than head) and reddish at
>   >   base in adults, head and throat mostly grayish
>   >   . . . Virginia rail.
>   cc. Bill over 2 inches (50 mm.) in length, larger birds
>   weighing more than 9 ounces (250 grams).
>   >   D. Head and upper parts of body mostly tinged
>   >   with ashy gray tones, found in salt-marsh
>   >   habitats . . . clapper rail.
>   >   DD. Head and upper parts of body mostly tinged
>   >   with reddish brown to buff tones, found in
>   >   fresh-water marsh habitats . . . king rail.

References

General

1. Amadon, D. 1943. Bird weights and egg weights. *Auk.* 60:221–34.
2. American Ornithologists' Union. 1957. *Check-list of North American birds.* 5th ed. Baltimore: Lord Baltimore Press.
3. Campbell, H., and Lee, L. 1953. *Studies on quail malaria in New Mexico and notes on other aspects of quail populations.* Santa Fe: New Mexico Department of Game and Fish.
4. Edminster, F. C. 1954. *American game birds of field and forest.* New York: Charles Scribner's Sons.
5. Giles, R. H., Jr. (ed.). 1969. *Wildlife management techniques.* Washington, D.C.: The Wildlife Society.
6. Irving, L. 1960. Birds of Anktuvuk Pass, Kobuk, and Old Crow: A study in Arctic adaptation. *U.S. National Museum Bulletin* 217:1–409.
7. Johnsgard, P. A. 1973. *Grouse and quails of North America.* Lincoln: University of Nebraska Press.
8. Johnson, R. E., and Lockner, J. R. 1968. Heart size and altitude in ptarmigan. *Condor* 70:185.
9. Johnston, D. W. 1963. Heart weights of some Alaska birds. *Wilson Bulletin* 75:435–46.
10. Miller, A. H., and Stebbins, R. C. 1964. *The lives of desert animals in Joshua Tree National Monument.* Berkeley: University of California Press.
11. *National survey of hunting and fishing.* 1972. U.S. Fish and Wildlife Service Resource Publication no. 95.
12. Nelson, A. L., and Martin, A. C. 1953. Gamebird weights. *Journal of Wildlife Management* 17:36–42.
13. Parmelee, D. F., Stephens, H. A., and Schmidt,

R. H. 1967. The birds of southeastern Victoria Island and adjacent small islands. *National Museum of Canada Bulletin* 222:1–229.

14. Petrides, G. A. 1942. Age determination in American gallinaceous game birds. *Transactions of the Seventh North American Wildlife Conference.* Pp. 308–28.

15. Robbins, C. S., and Van Velzen, W. T. 1969. *The breeding bird survey, 1967 and 1968.* U.S. Fish and Wildlife Service, Bureau of Sport Fisheries and Wildlife, Special Scientific Report (Wildlife) no. 124.

16. Tso-hsin, C. (ed.). 1963. [*China's economic fauna: Birds.* Peiping: Science Publishing Company]. Translated in 1964 by U.S. Department of Commerce, Washington, D.C.

References to Individual Species or Species Groups

American Coot

17. Fredrickson, L. H. 1968. Measurements of coots related to sex and age. *Journal of Wildlife Management* 32:409–11.

18. Frederickson, L. H. 1970. Breeding biology of American coots in Iowa. *Wilson Bulletin* 82:445–57.

19. Gullion, W. G. 1952a. Sex and age determination in the American coot. *Journal of Wildlife Management* 16:191–97.

20. Gullion, W. G. 1952b. The displays and calls of the American coot. *Wilson Bulletin* 64:83–97.

21. Gullion, W. G. 1954. The reproductive cycle of American coots in California, *Auk* 71:360–412.

22. Kiel, W. H. 1955. Nesting studies of the coot in southwestern Manitoba. *Journal of Wildlife Management* 19:189–98.

American Woodcock

23. Liscinsky, S. A. 1972. *The Pennsylvania woodcock management study*. Pennsylvania Game Commission Research Bulletin no. 171.
24. Martin, F. 1964. Woodcock sex and age determination from wings. *Journal of Wildlife Management* 28:287–93.
25. Sheldon, W. G. 1967. *The book of the American woodcock*. Amherst: University of Massachusetts Press.

Band-tailed Pigeon

26. Drewein, R. C., Vernimen, R., Harris, S., and Yocom, C. 1966. Spring weights of band-tailed pigeons. *Journal of Wildlife Management* 30:190–92.
27. MacGregor, W. G., and Smith. W. M. 1955. Nesting and reproduction of the band-tailed pigeon in California. *California Fish and Game* 41:315–26.
28. Peeters, H. J. 1962. Nuptial behavior of the band-tailed pigeon in the San Francisco area. *Condor* 64:445–70.
29. Smith, W. A. 1968. The band-tailed pigeon in California. *California Fish and Game* 54:1–16.

Blue Grouse

30. Boag, D. A. 1965. Indicators of sex, age, and breeding phenology in blue grouse. *Journal of Wildlife Management* 29:103–8.
31. Mussehl, T. W., and Leil, T. H. 1963. Sexing wings of adult blue grouse. *Journal of Wildlife Management* 27:102–6.

Bobwhite

32. Haugen, A. O. 1957. Distinguishing juvenile from adult bobwhite quail. *Journal of Wildlife Management* 21:29–32.
33. Thomas, K. P. 1969. Sex determination of bobwhites by wing criteria. *Journal of Wildlife Management* 33:215–16.

California Quail

34. Sumner, E. L. 1935. A life history study of the California quail, with recommendations for its conservation and management. *California Fish and Game* 21:167–253.

Chachalaca

35. Delacour, J., and Amadon, D. 1973. *Curassows and related birds.* New York: American Museum of Natural History.
36. Vaurie, C. 1965. Systematic notes on the bird family Cracidae. No. 2. Relationships and geographical distribution of *Ortalis vetula, Ortalis poliocephala and Ortalis leucogastra. American Museum Novitates* 2222:1–36.

Chukar Partridge

37. Bohl, Wayne. 1957. Chukars in New Mexico, 1931–1957. New Mexico Department of Game and Fish Bulletin no. 6.

Common Snipe

38. Tuck, L. M. 1972. *The snipes: A study of the genus Capella.* Canadian Wildlife Service, Ottawa.

Gallinules

39. Meanley, B. 1963. Pre-nesting activity of the purple gallinule near Savannah, Georgia. *Auk* 80:545–47.

Gambel Quail

40. Gullion, C. W. 1956. Let's go desert quail hunting. Nevada Fish and Game Commission Biological Bulletin no. 2.

Gray Partridge

41. McCabe, R. A., and Hawkins, A. S. 1946. The Hungarian partridge in Wisconsin. *American Midland Naturalist* 36:1–75.

Harlequin Quail

42. Leopold, A. S., and McCabe, R. A. 1957. Natural history of the Montezuma quail in Mexico. *Condor* 59:3–26.

Mountain Quail

43. McLean, D. D. 1930. *The quail of California.* California Division of Fish and Game, Game Bulletin no. 2.

Mourning Dove

44. Hanson, H. D., and Kossack, C. W. 1963. *The mourning dove in Illinois.* Illinois Department of Conservation Technical Bulletin no. 2.

Prairie Grouse (Pinnated and Sharp-tailed Grouse)

45. Ammann, G. A. 1944. Determining the age of pinnated and sharp-tailed grouse. *Journal of Wildlife Management* 8:170–71.
46. Ammann, G. A. 1957. *The prairie grouse of Michigan.* Michigan Department of Conservation Technical Bulletin.
47. Copelin, F. F. 1963. *The lesser prairie chicken in Oklahoma.* Oklahoma Wildlife Conservation Department Technical Bulletin no. 6.
48. Henderson, F. R., Brooks, F. W., Wood, R. E., and Dahlgren, R. B. 1967. Sexing of prairie grouse by crown feather patterns. *Journal of Wildlife Management* 31:764–69.
49. Lehmann, V. W. 1941. Attwater's prairie chicken: Its life history and management. U.S. Department of the Interior, Fish and Wildlife Service, *North American Fauna* 57:1–66.

Ptarmigans

50. Bergerud, A. T., Peters, S., and McGrath, R. 1963. Determining sex and age of willow ptarmigan in Newfoundland. *Journal of Wildlife Management* 27:700–711.
51. Braun, C. E. 1969. Population dynamics, habitat, and movements of white-tailed ptarmigan in Col-

orado. Ph.D. dissertation, Colorado State University.

52. Braun, C. E., and Rogers, C. E. 1967. *Determination of age and sex of the southern white-tailed ptarmigan.* Colorado Game, Fish and Parks Department Game Information Leaflet no. 54.

53. Weeden, R. B., and Watson, A. 1967. Determining the age of rock ptarmigan in Alaska and Scotland. *Journal of Wildlife Management* 31:825–26.

Rails

54. Adams, D. A., and Quay, T. L. 1958. Ecology of the clapper rail in southeastern North Carolina. *Journal of Wildlife Management* 22:149–56.

55. Horak, C. J. 1970. A comparative study of the foods of the sora and Virginia rail. *Wilson Bulletin* 82:206–13.

56. Kaufman, G. W. 1971. Behavior and ecology of the sora, *Porzana carolina,* and Virginia rail, *Rallus limicola.* Ph.D. dissertation, University of Minnesota, Minneapolis.

57. Mangold, R. E. 1973. Noisy phantom of the salt marsh: The clapper rail. *New Jersey Outdoors* 23(12):3–10.

58. Meanley, B. 1969. Natural history of the king rail. U.S. Fish and Wildlife Service, Bureau of Sport Fisheries and Wildlife, *North American Fauna* 67:1–108.

59. Pospichal, L. B., and Marshall, W. H. 1954. A field study of sora rail and Virginia rail in central Minnesota. *Flicker* 26:2–32.

60. Stewart, R. E. 1951. Clapper rail populations of the

Middle Atlantic States. *Transactions of the Sixteenth North American Wildlife Conference.* Pp. 421–30.

61. Walkinshaw, L. H. 1937. The Virginia rail in Michigan. *Auk* 54:464–75.
62. Walkinshaw, L. H. 1940. Summer life of the sora rail. *Auk* 57:153–68.

Ring-necked Pheasant

63. Baskett, T. S. 1947. Nesting and production of the ring-necked pheasant in north-central Iowa. *Ecological Monographs* 17:1–30.
64. Baxter, W. L., and Wolfe, C. W. 1973. *Life history and ecology of the ring-necked pheasant in Nebraska.* Nebraska Game and Parks Commission.

Ruffed Grouse

65. Bump, G., Darrow, R., Edminster, F., and Crissey, W. 1947. *The ruffed grouse: Life history, propagation, management.* Albany: New York State Conservation Department.
66. David, J. A. 1969. Aging and sexing criteria for Ohio ruffed grouse. *Journal of Wildlife Management* 33:628–36.
67. Dorney, R. S., and Holzer, F. V. 1957. Spring aging methods for ruffed grouse cocks. *Journal of Wildlife Management* 21:268–74.
68. Hale, J. B., Wendt, R. F., and Halazon, G. C. 1954. *Sex and age criteria for Wisconsin ruffed grouse.* Wisconsin Conservation Department Bulletin no. 9.

Sage Grouse

69. Eng, R. L. 1955. A method for obtaining sage grouse age and sex ratios from wings. *Journal of Wildlife Management* 19:267–72.
70. Patterson, R. L. 1952. *The sage grouse in Wyoming.* Denver: Sage Books.
71. Pyrah, D. B. 1963. *Sage grouse investigations.* Idaho Fish and Game Division Job Completion Report, Project W 125-R-2.

Sandhill Crane

72. Drewien, R. C. 1973. Ecology of Rocky Mountain greater sandhill cranes. Ph.D. dissertation, University of Idaho, Moscow.
73. Springs, A. J. 1962. A report on the 1961 sandhill crane season in Texas. Mimeo report, Texas Game and Fish Commission, Austin.
74. Walkinshaw, L. H. 1973. Cranes of the world. New York: Winchester Press.

Scaled Quail

75. Wallmo, O. C. 1956. Determination of sex and age of scaled quail. *Journal of Wildlife Management* 20:154–58.

Spruce Grouse

76. Ellison, L. N. 1968. Sexing and aging Alaskan

spruce grouse by plumage. *Journal of Wildlife Management* 32:12–16.

77. Stoneberg, R. P. 1967. A preliminary study of the breeding biology of the spruce grouse in northwestern Montana. Master's thesis, University of Montana.

78. Zwickel, F. C., and Martinsen, C. F. 1967. Determining age and sex of Franklin spruce grouse by tails alone. *Journal of Wildlife Management* 31:760–63.

Turkey

79. Hewitt, O. H. (ed.). 1967. *The wild turkey and its management.* Washington, D.C.: The Wildlife Society.

80. Schorger, A. W. 1966. *The wild turkey: Its history and domestication.* Norman: University of Oklahoma Press.

White-winged Dove

81. Cottam, C., and Trefethen, J. B. 1968. *White-wings: The life history, status and management of the white-winged dove.* Princeton: D. Van Nostrand Co.

Index

This index contains, in a single alphabetical sequence, common names (roman type), including the most commonly used alternative vernacular names; scientific names (italics); and ordinal names (boldface).